INLAND FISH DIVERSITY

ABOUT AUTHORS

H.S. Mogalekar is Ph.D. Research Scholar working on fish diversity and productivity of selected reservoirs of south Tamil Nadu in Department of Fisheries Biology and Resource Management, Fisheries College and Research Institute, Tamil Nadu Fisheries University, Thoothukudi, Tamil Nadu, India. He received B.F.Sc. from College of Fishery Science, Maharashtra Animal and Fisheries Science University, Nagpur. Maharashtra and M.F.Sc. from Kerala University of Fisheries and Ocean Studies, Panangad, Kochi, Kerala, India. He has published 13 research and review papers in the field of freshwater and estuarine fish diversity.

J. Canciyal (M.F.Sc.) is Ph.D. Research Scholar in Department of Fisheries Resource Management, Faculty of Fishery Sciences, West Bengal University of Animal and Fisheries Sciences, Kolkata, India. She has been selected as scientist by Agricultural Scientists Recruitment Board under Indian Council of Agricultural Research, New Delhi, India. She has published eight papers and three course manuals. Her field of specialization are freshwater fish diversity and seaweed taxonomy.

Dr. P. Jawahar (M.F.Sc., Ph.D.) is working as Professor in the Department of Fisheries Biology and Resource Management, Fisheries College and Research Institute, Tamil Nadu Fisheries University, Thoothukudi, Tamil Nadu, India. He handles courses like Taxonomy of Fishes, Shellfish Taxonomy, Fish population dynamics and stock assessment, Marine Fisheries Resource Management, Modern Techniques in Ichthyo-taxonomy, Tropical fish stock assessment and Application of Fisheries Models in Stock Assessment for under graduate and postgraduate students. He has published 73 research papers, 5 teaching manuals and 13 training manuals.

INLAND FISH DIVERSITY

H.S. Mogalekar
J. Canciyal
P. Jawahar

2017
Scholars World
A Division of
Astral International Pvt. Ltd.
New Delhi – 110 002

ISBN 9789387057661 (International Edition)

Published by : **Scholars World**
A Division of
Astral International Pvt. Ltd.
– ISO 9001:2015 Certified Company –
4760-61/23, Ansari Road, Darya Ganj
New Delhi-110 002
Ph. 011-4354 9197, 2327 8134
E-mail: info@astralint.com
Website: www.astralint.com

TAMIL NADU FISHERIES UNIVERSITY

Prof. Baskaran Manimaran
Vice-Chancellor

Nagapattinam - 611 001
Tamil Nadu, India

Foreword

South Indian states are blessed with rich freshwater resources with a diversified fish fauna characterized by many endemic fish species. The rivers originating from Western Ghats and Eastern Ghats harbour a number of endemic fishes suitable for aquarium keeping, sport fishing and aquaculture. There are many unexplored inland water bodies in south India which may harbour fish species yet to be discovered and documented, many inadequately known fish species to be re-described and decisions to be taken on protection of habitats based on their conservation status. In recent years, novel facts arising out of rigorous investigations of several inland fishes reformed the systematics of some of the Indian fishes. There is urgent need of scientific community to conduct intensive fish faunal surveys through field collections and literature analysis for appropriate documentation of the inland fishes.

There is immense scope for fish production, nutritional security and income generation through captive breeding and farming of native fish species in south Indian states. Unfortunately breeding and culture techniques of several native fish species are yet to be established. In order to increase inland fish production, fast growing exotic fishes and Gangetic carps were introduced in southern states which lead to total negligence of native fish fauna. The endemic cyprinids form most diverse group among the freshwater fishes in these eco-regions and many species form commercially important fisheries and aquaculture. In future, these native fish species could be utilised to increase fish production through aquaculture. Anthropogenic activities caused steady decline of native fish fauna in their natural habitats. Lack of adequate scientific information on native fish species has been a major hurdle in taking timely step in Conservation. In order to address these issues knowledge on taxonomy, environment, distribution and abundance in eco-region, food and feeding, maximum size, reproduction, fishery status and conservation status are essential.

This book on "Inland Fish Diversity of South India" is a concerted effort to document freshwater fishes of Karnataka, Kerala, Tamil Nadu and United Andhra Pradesh as well as estuarine fishes of Kerala and Tamil Nadu. Mr. Mogalekar, Miss. J. Canciyal and Dr. P. Jawahar deserve all the appreciation for bringing out this pioneering publication. Through this book, the authors tried to address the above issues comprehensively. I hope this publication will be an indispensable reference to students, researchers and resource managers in south India.

9/3/16

(Baskaran Manimaran)

Preface

The aquatic resources of Karnataka, Kerala, Tamil Nadu and United Andhra Pradesh are enriched with a vast expanse of freshwater fisheries resources. The Western Ghats Mountains along the west coastal line, Eastern Ghats to the east coastal line and the Deccan plateau with the Sahyadri range to the north border are three major mountain ranges constitute extremely important life supporting ecosystems in the South India. These south Indian states cover about 22.9% length of perennial rivers and 40.72% of reservoirs resources of India which provides immense scope for inland fisheries development in the country. The fast flowing streams and rivers have been an excellent habitat and environment enabling evolution of rich fish diversity. The ironic appearance of freshwater fish diversity with high degree of endemism is met in the streams and major rivers of south India.

In spite of series of publications on the freshwater fishes of south India, consolidated list of freshwater fishes endemic to these regions has not been published so far. Hence forward we made the best use of available information on inland fish fauna of south India in developing this publication. This publication reflects restructured information on freshwater fish diversity of Karnataka, Kerala, Tamil Nadu and United Andhra Pradesh and estuarine fish diversity of Tamil Nadu and finfish diversity of Vembanad Lake in the Panangad-Kumbalam region of Kerala with note on systematic placement as per the latest taxonomic revisions of fish species, together with their environment, endemic and exotic status, maximum size, human use (existence or nonexistence of fishery, aquaculture, ornamental or game fish) and conservation status as per IUCN criteria. Adequate attention has not been paid for the sustainable utilization and conservation of endemic freshwater fish fauna of south India. Many endemic freshwater fishes of south India have been listed under threatened category now either due to attempts to promote aquaculture practices using transplanted Indian major carps or other exotic species, over exploitation, dynamite fishing or construction of dams, canals or barrages across the rivers.

Generation of proper database on fish germplasm resources reveals crystal facts necessary for the management, conservation and to evolve a plan for sustainable utilization of fishery resources. We look forward for suggestions and constructive criticism from our enlightened readers for additional enhancement of this publication. It is hoped that this book will provide basic information on the inland fish diversity of south India and offer better inputs for scientific management and utilization of the resources on the sustainable manner.

We wish to express our deep sense of gratitude to the Dr. G. Sugumar, Dean, Fisheries College and Research Institute, Thoothukudi-628008, Tamil Nadu for necessary support provided to complete this work. The sincere appreciation and heartfelt thanks to Mr. R. Kumaresan for provided assistance and cooperation in the college library during book writing. We are deeply indebted to the persons contributed to inland fisheries research in south India.

H. S. Mogalekar
J. Canciyal
P. Jawahar

Contents

1

SUMMARY OF INLAND FISH DIVERSITY OF SOUTH INDIA

H.S. Mogalekar

Department of Fisheries Biology and Resource Management, Fisheries College and Research Institute, Tamil Nadu Fisheries University, Thoothukudi-628 008, Tamil Nadu, India

1. Freshwater fish diversity of Karnataka

The updated checklist of the freshwater fish diversity of Karnataka includes 245 species (164 freshwater, 65 freshwater cum brackish and 16 marine cum freshwater cum brackish) distributed under 103 genera, 39 families and 14 orders. Among these, 154 species are endemic to Indian sub-continent, 74 species are endemic to India, 57 species are endemic to Western Ghats of India and 11 species are exotic. The most diverse order was cypriniformes with 144 species, 45 genera and 5 families, followed by siluriformes (51 species, 24 genera and 9 families) and perciformes (27 species, 16 genera and 10 families). Top four families with the maximum number of species were cyprinidae (123 species), bagridae (20), nemacheilidae (15) and schilbeidae (10). Forty species listed under threatened category including 6 species as critically endangered, 21 as endangered and 14 as vulnerable. As numbers of commercial, sports, ornamental and cultivable fishes are high, commercial and recreational fishing could be organized and seed production of native species by selective breeding is recommended for aquaculture practices.

2. Freshwater fish diversity of Kerala

A systematic, updated checklist of freshwater fish species of Kerala comprises of 273 species of fishes (196 freshwater species and 77 diadromous species) belonging to 16 orders, 45 families and 106 genera. Freshwater fish species of Kerala consist of 113 species with ornamental value, 44 species forms candidate species for aquaculture, 119 species are important for capture fishery and 26 species are important for sport fishing. The top order with diverse species composition was Cypriniformes with 154 species, 44 genera and 4 families. Most diverse family was Cyprinidae with total of 121 species from 32 genera. Sixty species were found to be confined to the freshwater bodies of Kerala. Among the species listed under threatened category, 6 were critically endangered while 44 species were endangered, whereas 21 species were vulnerable. Enduring management plans are obligatory to conserve fish germplasm as 26.0% of fishes are coming under threatened category.

3. Freshwater fish diversity of Tamil Nadu

Tamil Nadu is bestowed with diverse freshwater fish germplasm consists of 218 species (including 125 were primary freshwater fishes and 93 saltwater dispersents) representing 95 generas, 35 families and 14 orders. Cyprinidae contributed 48.16 % to total fish fauna with 105 species belonging to 32 genera, followed by Bagridae (13 species from three genera), Gobiidae (10 species from nine genera), Cichlidae (seven species from four genera), Ambassidae (seven species from three genera) and Nemacheilidae (seven species from two genera). 15 species are endemic to Tamil Nadu. Fish species with ornamental, food and ornamental as well as food are 139, 114 and 35 in numbers respectively. Based on the information collected on conservation status of all taxa, four species listed as Critically Endangered, 20 as Endangered, seven as Vulnerable and 10 as Near Threatened. Appropriate conservation strategies need to be proposed to overcome the problemes of species endangerment.

4. Freshwater fish diversity of United Andhra Pradesh

The updated checklist of freshwater fish diversity of United Andhra Pradesh revealed occurrence of 227 species of fishes (136 primary freshwater species and 91 diadromous species) belonging to 108 genera in 37 families and 12 orders. These consist of 79 species with ornamental value, 51 species forms candidate species for aquaculture, 134 species are important for capture fishery and 25 species are important for game fishing / sport fishing. Of the 227 species, 16 species are exotic and alien, 5 species are endemic to United Andhra Pradesh, 2 species are endemic to Maharashtra, 18 species are endemic to Western Ghats of India, 7 species are endemic to Peninsular India and 42 species are endemic to India. Among the species listed under threatened category, 3 species were critically endangered, 11 species were endangered and 6 species were vulnerable.

5. Estuarine fish diversity of Tamil Nadu

There are 38 estuaries known from Tamil Nadu of which 12 estuaries explored for fish biodiversity assessment. The checklist of estuarine fishes of Tamil Nadu contains 241 species distributed under 154 genera, 75 families and 17 orders. The most diverse order was perciformes with 126 species, 77 genera and 38 families. The top five families with the highest number of species were carangidae (18 species), cyprinidae (15), gobidae (14), engraulidae (13) and clupeidae (12). IUCN red list conservation status of all taxa includes 3 species as vulnerable, 12 near threatened,

1 lower risk-least concern, 65 least concern and 7 data deficient. As numbers of commercial, sports, ornamental and cultivable fishes are high, commercial and recreational fishing could be organized.

6. Finfish diversity of Vembanad Lake in the Panangad-Kumbalam region of Kerala

Studies on finfish diversity of Panangad-Kumbalam backwater revealed presence of 38 species of finfishes belonging to 27 families 11 orders. Based on the information collected on conservation status of all finfish taxa, three species listed as Near Threatened, 18 as Least Concern, two as Data Deficient and 15 species have not yet been evaluated for their conservation status by IUCN Red list. Representation of Cichlidae and Mugilidae was higher than other families. Recreational fishing may be organised in the Panangad-Kumbalam backwater as this region is endowed with food, ornamental and sport fishes.

2

FRESHWATER FISH DIVERSITY OF KARNATAKA

H.S. Mogalekar[1], *P. Jawahar*[1], *Prateek*[1] *and J. Canciyal*[2]

[1]*Department of Fisheries Biology and Resource Management, Fisheries College and Research Institute, Tamil Nadu Fisheries University, Thoothukudi-628 008, Tamil Nadu, India*

[2]*Department of Fisheries Resource Management, Faculty of Fishery Sciences, West Bengal University of Animal and Fisheries Sciences, Kolkata - 700 094, India*

INTRODUCTION

Karnataka is one of the richest Indian state in abundance of inland water resources. The State has 5.65 lakh ha of inland water resources, comprising 1.72 lakh ha of Departmental Tanks (3399 no.), 1.21 lakh ha of Gram Panchayat tanks (22624 no.), 2.72 lakh ha of reservoirs (82 no.) besides 5813 km length of rivers, 3187 km length of canals and 2000 ha of private fish culture ponds (Directorate of Fisheries in Karnataka, 2015). Rivers originating from Western Ghats conferred with diverse fish germplasm resources (Johnson and Arunachalam, 2009). Sharavati, Kali, Netravati, Bedthi, Gangavalli, Aghanashini, Varahi and Chakra are west flowing rivers in Karnataka. The tributaries of Krishna (Tungabhadra, Ghataprabha, Malaprabha, Bhima and Vedavati), Cauvery and Godavari (Hemavati, Kabini, Arkavati, Shimsha, Palar, Uttara and Dakshina Pinakini, Manjira and Karanja) form east flowing rivers. These water resources provide immense scope for inland fisheries development with estimated annual fish potential of around 4.01 lakh metric tons (Directorate of Fisheries in Karnataka, 2015).

The fish fauna inhabiting the freshwater ecosystems of Karnataka are diverse and fairly well known. Significant contributions to systematics of freshwater fish fauna of Karnataka were that of Chandrashekhariah *et al.* (2000); Jen (2002); Thippeswamy *et al.* (2008); Abida *et al.* (2009); Vijaylaxmi *et al.* (2010); Thirumala *et al.* (2011); Vijaylaxmi and Vijaykumar (2011); Ramachandra *et al.* (2012); Shahnawaz and Venkateshwarlu (2012); Shivashankar and Venkataramana (2012); Kumar (2013); Naik *et al.* (2013a); Naik *et al.* (2013b); Naik *et al.* (2013c); Naik *et al.* (2013d); Anandhi and Sharath (2014); Bhat and Hegde (2014); Naik *et al.* (2014); Venkateshwarlu *et al.* (2014); Britz and Ali (2015). Since publication of these pioneering works, numerous changes in taxonomy and systematic placement of species took place and comprehensive review on the entire freshwater fish fauna of Karnataka is not available now. The list of species presented earlier by many authors is updated by inclusion of all species of freshwater fishes now known to occur in the Karnataka. Henceforward present checklist is intended to update the scientific nomenclature of all the fish species recorded from freshwater bodies of Karnataka to assist fish taxonomists in future systematic work.

MATERIALS AND METHODS

The present checklist is based on the available published literature present in the form of research articles, monographs, books, species checklists and technical reports. Jayaram (2010), Fishbase (Froese and Pauly, 2013) and Catalog of fishes by Eschmeyer (Eschmeyer *et al.*, 2016) have been referred for confirmation and updating the orders, families, generas and species. Information on the environment, endemic or exotic status and human use (existence or non existence of fishery, aquaculture, ornamental or gamefish) of all species was obtained by retrieving species level information from the published literature as well as FishBase (Froese and Pauly, 2013). The endemicity or exoticity of all species indicated based on Indian subcontinent. The list of ornamental fishes is prepared based on colouration pattern, shape and maximum size. Information on the conservation status of all taxa in this paper was retrieved from the International Union for Conservation of Nature (IUCN) red list of threatened species, downloaded on 27 January 2016 (IUCN, 2015).

RESULTS

Systematic classification of fish species recorded in the basins of Cauvery, Krishna, Godavari, West-flowing rivers, reservoirs and Irrigation tanks of the Karnataka are given in Table 1. The consolidated checklist of freshwater fishes of Karnataka includes a total of 245 species, which were distributed among 103 genera, 39 families and 14 orders (Table 2). Among these, 163 species were freshwater, 65 were freshwater cum brackish and the remaining 16 were marine cum freshwater cum brackish (Table 1). The top three orders with diverse species composition were Cypriniformes (144 species, 45 genera and 5 families), Siluriformes (51 species, 24 genera and 9 families) and Perciformes (27 species, 16 genera and 10 families). The most diverse family was the Cyprinidae with 123 species and 35 genera, followed by Bagridae with 20 species and 6 genera, Nemacheilidae with 15 species and 5 genera, Schilbeidae with 10 species and 7 genera (Table 2 & Figure 1).

Of the 245 species, 39 species are distributed all over India, 10 species are endemic to Peninsular India, 4 species are endemic to south India, 30 species are endemic to Western Ghats of India, 9 species are endemic to Cauvery River, 5 species are endemic to Krishna River, one species is endemic to Sharavati River, 4 species are endemic to Maharashtra, 14 species are endemic to Karnataka, 9 species are endemic to Kerala, 2 species are endemic to Tamil Nadu and 15 species are exotic and alien (Table 1). *Garra bicornuta* (Narayan Rao, 1920), *Schismatorhynchos nukta* (Sykes, 1839) and *Nemacheilus semiarmatus* Day, 1867 are endemic to Maharashtra and Karnataka. *Puntius thomassi* (Day, 1874), *Longischistura striata* (Day, 1867), *Parambassis thomassi* (Day, 1870) and *Pterocryptis wynaadensis* (Day, 1873) are endemic to Karnataka and Kerala. *Mystus oculatus* (Valenciennes, 1840) is endemic to Maharashtra and Kerala. *Proeutropiichthys taakree* (Sykes, 1839) is endemic to Kerala and Tamil Nadu.

A list of freshwater fishes of Karnataka comprises of 75 species with ornamental value, 48 species are being used for aquaculture, 54 species for capture fishery and 26 species for gamefishing. Among the species listed under threatened category, 6 species are critically endangered while 21 species are endangered, whereas 14 species are vulnerable. There are 205 species under the non-threatened category, among which 18 are near threatened, whereas 157 species belonged to least concern, 12 species belonged to data deficient and 17 species have not been evaluated against IUCN criteria (Table 1).

Species listed as Critically Endangered are *Barbodes bovanicus, Barbodes wynaadensis, Hypselobarbus pulchellus, Puntius deccanensis, Puntius deccanensis* and *Hemibagrus punctatus*. The endangered category includes *Botia striata, Barilius canarensis, Dawkinsia arulius, Garra hughi, Hypselobarbus curmuca, Hypselobarbus dobsoni, Hypselobarbus dubius, Hypselobarbus micropogon, Hypselobarbus mussullah, Labeo potail, Puntius cauveriensis, Schismatorhynchos nukta, Thynnichthys sandkhol, Tor khudree, Schistura nagodiensis, Etroplus canarensis, Batasio sharavatiensis, Pseudeutropius mitchelli, Silonia childreni, Pterocryptis wynaadensis* and *Glyptothorax madraspatanus.*

DISCUSSION

The India has rich freshwater fish genetic resources constituting 942 species which have been recorded in the FishBase database (Froese and Pauly, 2013). Karnataka constitute about 26 % to the freshwater fish diversity of India. Many species could be distributed in the drainage of the Western Ghats and other unexplored areas and therefore more biodiversity explorations are required. Chandrashekhariah *et al.* (2000) listed 169 species of freshwater fishes from the basins of Cauvery (97 species belonging to 23 families and 8 orders), Krishna (101 species belonging to 19 families and 5 orders), Godavari (60 species belonging to 13 families and 4 orders) and West-flowing rivers (84 species belonging 19 families and 7 orders) of the Karnataka. However, present record of 245 species is relatively higher compared to Chandrashekhariah *et al.* (2000). Existing conservation policies need to be implementd to reduce species extinction, as 40 species of freshwater fishes listed under threatened category. As numbers of ornamental as well as food fishes are high, seed production by selective breeding is recommended for income generation through culture of these species. Selective breeding and ranching of native fish species might help to overcome the problems of species endangerment.

Since the publication of Chandrashekhariah *et al.* (2000), 76 species of fishes has been reported from freshwater bodies of Karnataka by various authors. Jen (2002) listed 88 fish species from Cauvery in Karnataka. Vijaylaxmi and Vijaykumar (2011) reported 29 fish species belonging to 6 orders, 8 families and 18 genera from the Bheema River in Gulbarga District of Karnataka. Ramachandra *et al.* (2012) documented 64 species under 32 genera and 16 families from upper catchment Linganamakki Reservoir in the Sharavathi river basin of Karnataka. Shahnawaz and Venkateshwarlu (2012) recorded 48 species of fishes belonging to 3 families under the order cypriniformes from Tunga and Bhadra rivers of Karnataka. Shivashankar and Venkataramana (2012) documented 48 fish species under 28 genera, 14 families from Bhadra River in Shimoga District of Karnataka. Naik *et al.* (2013a) observed 32 species of finfishes belonging to 26 genera, 14 families and 6 orders from Upper Mullamari Reservoir of Karnataka. Naik *et al.* (2013b) listed 37 species of fishes belonging to 11 families and 4 orders from Tunga River of Karnataka. Naik *et al.* (2013c) recorded 78 fish species belonging to 18 families and 6 orders from Varada River of Karnataka. Naik *et al.* (2013d) observed 64 species of finfishes belonging to 37 genera, 16 families and 5 orders from the Karanja Reservoir of Karnataka. Naik *et al.* (2014) recorded 5 species of fishes belonging to 15 families and 5 orders from Chulkinala reservoir in north Karnataka. Venkateshwarlu *et al.* (2014) listed 82 species of fishes belonging to 42 genera, 20 families and 8 orders from Sita, Swarna and Varahi rivers in Udupi District of Western ghats of Karnataka. All the above reports agreed with dominance of cyprinids over other freshwater fish families. The Cyprinidae is the most diverse group among the freshwater fishes and many species form commercially important fisheries and most of interest to aquaculture.

The available literature pointed out the distressing impact of introduced species on the indigenous fish components in different freshwater bodies of Karnataka. Unintentional stocking of tilapia, exotic carps and Indian major carps has led to replacement of many endemic fish species like *Etroplus canarensis, Barbodes carnaticus, Hypselobarbus jerdoni, Hypselobarbus pulchellus, Labeo calbasu, Clarias dussumieri* and *Pangasius pangasius* (Chandrashekhariah *et al.*, 2000). These have been established themselves in most of the reservoirs by natural breeding. The construction of 65 reservoirs across the rivers Krishna, Godavari, Cauvery and west-flowing river basins in Karnataka has also affected the fish fauna to a considerable extent, especially, the migratory ones.

CONCLUSIONS

Karnataka is endowed with rich freshwater fish diversity. Proper conservation strategies need to be proposed as 16.32 % of endemic species facing the threat of endangerment. Exotic as well as transplanted species are well established in most of the freshwater systems, necessary preventive measures prerequisite to adopt. Daming and channelization across the rivers and streams affected on upstream and downstream migration of migratory species, proper migrating passages requisite to provided.

REFERENCES

Abida, B., Harikrishna, S. and Khan, I. 2009. Analysis of Heavy metals in Water, Sediments and Fish samples of Madivala Lakes of Bangalore, Karnataka. International Journal of Chem. Tech. Research 1(2): 245-249.

Anandhi, D.U. and Sharath, Y.G. 2014. Endemic Ornamental Fishes of Adda Hole, Kabbinale forest with reference to Western Ghats, Karnataka. Journal of Aquatic Biology and Fisheries 2: 795-798.

Bhat, S. S and Hegde, A.K. 2014. A Note on Fresh water Fish diversity in major Tributaries of River Bedti of Western Ghats region of Karnataka, India. Research Journal of Animal, Veterinary and Fishery Sciences 2(8): 5-10.

Britz, R. and Ali, A. 2015. Dario huli, a new species of badid from Karnataka, southern India (Teleostei: Percomorpha: Badidae). Zootaxa 3911(1): 139–144.

Chandrashekhariah, H.N., Rahman, M.F. and Lakshmi, R.S. 2000. Status of fish fauna in Karnataka. pp. 98-135. In: Ponniah, A.G. and Gopalakrishnan, A. (eds.), Endemic Fish Diversity of Western Ghats. NBFGR-NATP Publication. National Bureau of Fish Genetic Resources, Lucknow, U.P., India. 347pp.

Directorate of Fisheries in Karnataka. 2015. Karnataka fisheries at a glance. Department of Fisheries, Government of Karnataka, 2 p. www.karnataka.gov.in/fisheries/Pages/Publications

Eschmeyer, W. N., Fricke, R. and van der Laan, R. (eds). 2016. Catalog of fishes: genera, species, references. Electronic version accessed 19-28 March 2016. (http://researcharchive.calacademy.org/research/ichthyology/catalog/fishcatmain.asp). [This version was edited by Bill Eschmeyer.]

Froese, R. and Pauly, D. (Eds.) 2013. List of Freshwater Fishes reported from India. In: FishBase. www.fishbase.org, 23 October 2015.

IUCN, 2015. The IUCN Red List of Threatened Species. Version2015.2. IUCN, Gland, Switzerland and Cambridge, UK, http://www.iucnredlist.org. 23 October 2015.

Jayaram, K.C. 2010. The Freshwater Fishes of the Indian Region. 2nd Edition, Narendra Publishing House, New Delhi, 616 pp.

Jen, W. 2002. Freshwater Fish Species in Cauvery River (Karnataka, India). In: Froese, R. and Pauly, D. (eds) (2013). FishBase. World Wide Web electronic publication. http://fish.mongabay.com/data/ecosystems/Cavally%20River.htm. 23 October 2015.

Johnson, J.A. and Arunachalam, M. 2009. Diversity, distribution and assemblage structure of fishes in streams of southern Western Ghats, India. Journal of Threatened Taxa 1(10): 507-513.

Kumar, K.H. 2013. Limnological characteristics of Jannapura tank in Bhadravathi Taluk of Karnataka, India. Weekly Science Research Journal 2(46): 1-10.

Naik, A.S.K., Somashekara, S.R., Jitendra, K., Mahesh, V., Benakappa, S., Anjaneyappa, H.N. and Nayana, P. 2013a. Assessment of Fish Biodiversity in Upper Mullamari Reservoir, Basavakalyan, Karnataka (India). International Journal of Fisheries and Aquaculture Sciences 3(1): 13-20.

Naik, A.S.K., Jitendra, K., Mahesh, V. and Benakappa, S. 2013b. Assessment of Fish diversity of Tunga River, Karnataka, India. Nature and Science 11(2): 82-87.

Naik, A.S.K., Jitendra, K., Mahesh, V., Anjaneyappa, H.N., Benakappa, S., Somashekara, S.R., Rajanna, K.B. and Hulkoti, S.H. 2013c. Status of Fish Diversity of Varada River, Karnataka, India. Environment and Ecology 31(2A): 786-792.

Naik, A.S.K., Benakappa, S., Somashekara, S.R., Anjaneyappa, H.N., Jitendra, K., Mahesh, V., Hulkoti, S.H. and Rajanna, K.B. 2013d. Studies on Ichthyofaunal Diversity of Karanja Reservoir, Karnataka, India. International Research Journal of Environment Sciences 2(2): 1-5.

Naik, A.S.K., Jitendra, K., Benakappa, S., Somashekara, S.R., Anjanayappa, H.N., Manjappa, N. and Mahesh, V. 2014. Ichthyofaunal diversity of Chulkinala Reservoir. Animal Science Reporter 8(2): 48-60.

Ramachandra, T.V., Chandran, M.D.S., Joshi, N.V., Sreekantha, Raushan, K., Rajinikanth, R., Desai, S.R. and Subhash, B. 2012. Freshwater Fish Diversity and Fisheries in Sharavathi River Basin. In: Ecological Profile of Sharavathi River Basin. Sahyadri Conservation Series: 22, ENVIS Technical Report: 52.

Shahnawaz, A. and Venkateshwarlu, M. 2012. Habitat ecology of Cyprind fish communinty in relation to environmental factors of Tunga and Bhadr Rivers, Western Ghats, Karnataka (India). Journal of Research and Devolelopment 12: 65-91.

Shivashankar, P. and Venkataramana, G. V. 2012. Ichthyodiversity status with relation to water quality of Bhadra River, Western Ghats, Shimoga District, Karnataka. Annals of Biological Research 3(10): 4893-4903.

Talwar, P.K. and Jhingran, A.G. 1991. Inland Fishes of India and Adjacent Countries. Vol 1 and 2, Oxford and IBH Publishing Co., New Delhi, 1–1158 pp.

Thippeswamy, S., Sequeira, M.G. and Suresh, G.C. 2008. Hydrobiology of Kodige Tank in the Western Ghats, India. Proceedings of Taal: The 12th World Lake Conference; 12-18.

Thirumala, S., Kiran, B.R. and Kantaraj.G.S. 2011. Fish diversity in relation to physico-chemical characteristics of Bhadra reservoir of Karnataka, India. Advances in Applied Science Research 2(5): 34-47.

Venkateshwarlu, M., Shetty, B.A.K. and Kiran, B.R. 2014. Conservation status of fish diversity of rivers - Sita, Swarna and Varahi in Udupi District, Western Ghats, Karnataka, India. International journal of advanced scientific and technical research 4(1): 797-813.

Vijaylaxmi, C. and Vijaykumar, K. 2011. Biodiversity of fish fauna of the Bheema River in Gulbarga district of Karnataka. The Ecoscan 5(1and2): 21-25.

Vijaylaxmi, C., Rajshekhar, M. and Vijaykumar, K. 2010. Freshwater fishes distribution and diversity status of Mullameri River, a minor tributary of Bheema River of Gulbarga District, Karnataka. International Journal of Systems Biology 2(2): 01-09.

Table 1. Freshwater Fishes of Karnataka with note on environment, endemic or exotic status, human use, maximum size and conservation status

Order, Family, Scientific name	Environment	Endemic, Exotic	Human Use	Maximum size (TL)	IUCN Status	References
Order: Anguilliformes						
Family :Anguillidae						
Anguilla bengalensis (Gray, 1831)	Marine; freshwater; brackish		Aquaculture; Gamefish	200 cm	NT	Chandrashekhariah *et al.* (2000); Jen (2002)
Anguilla bicolor McClelland, 1844	Marine; freshwater; brackish		Capture Fishery	123 cm	NT	Chandrashekhariah *et al.* (2000); Jen (2002)
Order: Beloniformes						
Family : Adrianichthyidae						
Oryzias melastigma (McClelland, 1839)	Freshwater; brackish		Fishery doesn't exists	4.0 cm	LC	Chandrashekhariah *et al.* (2000)
Oryzias setnai (Kulkarni, 1940)	Freshwater; brackish	ENSI	Ornamental	3.0 cm	LC	Chandrashekhariah *et al.* (2000)
Family : Belonidae						
Strongylura strongylura (van Hasselt, 1823)	Marine; freshwater; brackish		Capture Fishery	40.0 cm SL	NE	Chandrashekhariah *et al.* (2000)
Xenentodon cancila (Hamilton, 1822)	Marine; freshwater; brackish		Ornamental	40.0 cm	LC	Chandrashekhariah *et al.* (2000); Vijaylaxmi *et al.* (2010); Vijaylaxmi & Vijaykumar (2011); Shivashankar & Venkataramana (2012)

Order, Family, Scientific name	Environment	Endemic, Exotic	Human Use	Maximum size (TL)	IUCN Status	References
Family: Hemiramphidae						
Hyporhamphus limbatus (Valenciennes, 1847)	Marine; freshwater; brackish		Capture Fishery	35.0 cm	LC	Chandrashekhariah *et al.* (2000)
Hyporhamphus xanthopterus (Valenciennes, 1847)	Marine; freshwater; brackish		Ornamental	15.0 cm	VU	Venkateshwarlu *et al.* (2014)
Order: Clupiformes						
Family: Clupeidae						
Tenualosa ilisha (Hamilton, 1822)	Marine; freshwater; brackish		Aquaculture	42 cm	LC	Jen (2002)
Family: Engraulidae						
Setipinna phasa (Hamilton, 1822)	Freshwater; brackish	ENI	Capture Fishery	40.0 cm	LC	Naik *et al.* (2013a); Naik *et al.* (2013c)
Order: Cypriniformes						
Family: Balitoridae						
Bhavania australis (Jerdon, 1849)	Freshwater	ENWG	Ornamental	9.0 cm SL	LC	Chandrashekhariah *et al.* (2000); Anandhi & Sharath (2014); Venkateshwarlu *et al.* (2014)

Order, Family, Scientific name	Environment	Endemic, Exotic	Human Use	Maximum size (TL)	IUCN Status	References
Balitora mysorensis Hora, 1941	Freshwater	ENWG	Fishery doesn't exists	5.0 cm SL	VU	Chandrashekhariah *et al.* (2000); Shahnawaz & Venkateshwarlu (2012); Shivashankar & Venkataramana (2012); Venkateshwarlu *et al.* (2014)
Family:Cobitidae						
Botia almorhae Gray, 1831	Freshwater	ENI	Fishery doesn't exists	15.5 cm SL	LC	Naik *et al.* (2013c); Naik *et al.* (2013d)
Botia striata Narayan Rao, 1920	Freshwater	ENWG	Ornamental	7.8 cm SL	EN	Chandrashekhariah *et al.* (2000); Shivashankar & Venkataramana (2012)
Lepidocephalichthys thermalis (Valenciennes, 1846)	Freshwater		Ornamental	38.0 cm SL	LC	Chandrashekhariah *et al.* (2000); Jen (2002); Shahnawaz & Venkateshwarlu (2012); Ramachandra *et al.* (2012)
Family: Cyprinidae						
Amblypharyngodon melettinus (Valenciennes, 1844)	Freshwater		Ornamental	8.0 cm	LC	Chandrashekhariah *et al.* (2000); Ramachandra *et al.* (2012); Kumar (2013)
Amblypharyngodon mola (Hamilton, 1822)	Freshwater		Fishery doesn't exists	20.0 cm	LC	Chandrashekhariah *et al.* (2000); Abida *et al.* (2009); Shahnawaz & Venkateshwarlu (2012); Naik *et al.* (2013c); Naik *et al.* (2013d)
Bangana ariza (Hamilton, 1807)	Freshwater	ENI	Capture Fishery	30.0 cm SL	LC	Chandrashekhariah *et al.* (2000); Jen (2002)

Order, Family, Scientific name	Environment	Endemic, Exotic	Human Use	Maximum size (TL)	IUCN Status	References
Bangana dero (Hamilton, 1822)	Freshwater		Capture Fishery	75.0 cm	LC	Jen (2002)
Barbodes bovanicus (Day, 1877)	Freshwater	ENCR	Fishery doesn't exists	36.0 cm	CR	Chandrashekhariah *et al.* (2000); Jen (2002)
Barbodes carnaticus (Jerdon, 1849)	Freshwater	ENWG	Aquaculture	60.0 cm	LC	Chandrashekhariah *et al.* (2000); Jen (2002); Shivashankar & Venkataramana (2012); Naik *et al.* (2013c)
Barbodes wynaadensis (Day, 1873)	Freshwater	ENI	Capture Fishery	25.0 cm	CR	Chandrashekhariah *et al.* (2000); Jen (2002)
Barilius bakeri Day, 1865	Freshwater	ENWG	Ornamental	15.0 cm	LC	Chandrashekhariah *et al.* (2000); Ramachandra *et al.* (2012); Anandhi & Sharath (2014); Venkateshwarlu *et al.* (2014)
Barilius barila (Hamilton, 1822)	Freshwater		Ornamental	10.0 cm	LC	Chandrashekhariah *et al.* (2000)
Barilius barna (Hamilton, 1822)	Freshwater		Capture Fishery	15.0 cm	LC	Chandrashekhariah *et al.* (2000); Naik *et al.* (2013a); Naik *et al.* (2013c); Naik *et al.* (2013d); Venkateshwarlu *et al.* (2014)
Barilius bendelisis (Hamilton, 1807)	Freshwater		Capture Fishery	22.7 cm	LC	Chandrashekhariah *et al.* (2000); Vijaylaxmi *et al.* (2010); Vijaylaxmi & Vijaykumar (2011); Shahnawaz & Venkateshwarlu (2012); Shivashankar & Venkataramana (2012)

Order, Family, Scientific name	Environment	Endemic, Exotic	Human Use	Maximum size (TL)	IUCN Status	References
Barilius canarensis (Jerdon, 1849)	Freshwater	ENI	Ornamental	15.0 cm	EN	Chandrashekhariah *et al.* (2000); Thippeswamy *et al.* (2008); Shahnawaz & Venkateshwarlu (2012); Ramachandra *et al.* (2012); Venkateshwarlu *et al.* (2014)
Barilius gatensis (Valenciennes, 1844)	Freshwater	ENWG	Ornamental	15.0 cm	LC	Chandrashekhariah *et al.* (2000); Jen (2002); Ramachandra *et al.* (2012); Shahnawaz & Venkateshwarlu (2012); Shivashankar & Venkataramana (2012)
Cabdio morar (Hamilton, 1822)	Freshwater		Fishery doesn't exists	20.0 cm	LC	Chandrashekhariah *et al.* (2000); Naik *et al.* (2013c)
Catla catla (Hamilton, 1822)	Freshwater; brackish		Aquaculture; Gamefish	182 cm	LC	Chandrashekhariah *et al.* (2000); Vijaylaxmi & Vijaykumar (2011); Naik *et al.* (2013a); Naik *et al.* (2013b)
Chela cachius (Hamilton, 1822)	Freshwater; brackish		Ornamental	6.0 cm	LC	Chandrashekhariah *et al.* (2000)
Cirrhinus cirrhosus (Bloch, 1795)	Freshwater; brackish	ENI	Aquaculture; Gamefish	100.0 cm SL	VU	Chandrashekhariah *et al.* (2000); Naik *et al.* (2013c); Naik *et al.* (2013d); Naik *et al.* (2014)
Cirrhinus fulungee (Sykes, 1839)	Freshwater	ENPI	Capture Fishery	30.0 cm SL	LC	Chandrashekhariah *et al.* (2000); Thirumala *et al.* (2011); Shahnawaz & Venkateshwarlu (2012); Shivashankar & Venkataramana (2012); Kumar (2013)

Order, Family, Scientific name	Environment	Endemic, Exotic	Human Use	Maximum size (TL)	IUCN Status	References
Cirrhinus mrigala (Hamilton, 1822)	Freshwater		Fishery doesn't exists	99.0 cm	LC	Chandrashekhariah *et al.* (2000); Vijaylaxmi & Vijaykumar (2011); Naik *et al.* (2013a); Naik *et al.* (2013b); Naik *et al.* (2013c); Naik *et al.* (2013d)
Cirrhinus reba (Hamilton, 1822)	Freshwater		Fishery doesn't exists	30.0 cm	LC	Chandrashekhariah *et al.* (2000); Thippeswamy *et al.* (2008); Shahnawaz & Venkateshwarlu (2012); Shivashankar & Venkataramana (2012); Naik *et al.* (2013c); Naik *et al.* (2013d)
Ctenopharyngodon idella (Valenciennes, 1844)	Freshwater	EX	Aquaculture; Gamefish	150 cm	NE	Chandrashekhariah *et al.* (2000); Shivashankar & Venkataramana (2012); Kumar (2013); Naik *et al.* (2013a); Naik *et al.* (2013b); Naik *et al.* (2013c); Naik *et al.* (2013d); Naik *et al.* (2014)
Cyprinus carpio var. communis (Linnaeus, 1758)	Freshwater; brackish	EX	Aquaculture; Gamefish; Ornamental	120 cm	VU	Chandrashekhariah *et al.* (2000); Shahnawaz & Venkateshwarlu (2012); Kumar (2013); Naik *et al.* (2013a); Naik *et al.* (2013b); Naik *et al.* (2013c); Naik *et al.* (2013d); Naik *et al.* (2014)
Cyprinus carpio var. nudus (Bloch, 1784)	Freshwater	EX	Aquaculture	-	NE	Naik *et al.* (2013b); Naik *et al.* (2013c); Naik *et al.* (2013d); Naik *et al.* (2014)

Order, Family, Scientific name	Environment	Endemic, Exotic	Human Use	Maximum size (TL)	IUCN Status	References
Cyprinus carpio var. specularis (Lacepède, 1803)	Freshwater	EX	Aquaculture	33.0 cm	NE	Naik *et al.* (2013b); Naik *et al.* (2013c); Naik *et al.* (2013d); Naik *et al.* (2014)
Danio rerio (Hamilton, 1822)	Freshwater		Ornamental	3.8 cm SL	LC	Chandrashekhariah *et al.* (2000); Shahnawaz & Venkateshwarlu (2012); Ramachandra *et al.* (2012); Naik *et al.* (2013a)
Devario aequipinnatus (McClelland, 1839)	Freshwater		Ornamental	15.0 cm	LC	Chandrashekhariah *et al.* (2000); Shahnawaz & Venkateshwarlu (2012); Shivashankar & Venkataramana (2012); Ramachandra *et al.* (2012); Bhat & Hegde (2014)
Devario devario (Hamilton, 1822)	Freshwater		Ornamental	10.0 cm	LC	Chandrashekhariah *et al.* (2000)
Devario fraseri (Hora, 1935)	Freshwater	ENI	Ornamental	10.0 cm	LC	Chandrashekhariah *et al.* (2000)
Devario malabaricus (Jerdon, 1849)	Freshwater		Ornamental	12.0 cm	LC	Shahnawaz & Venkateshwarlu (2012); Shivashankar & Venkataramana (2012); Anandhi & Sharath (2014); Venkateshwarlu *et al.* (2014)
Devario annandalei (Chaudhuri, 1908)	Freshwater	EX	Fishery doesn't exists	9.0 cm SL	DD	Venkateshwarlu *et al.* (2014)

Order, Family, Scientific name	Environment	Endemic, Exotic	Human Use	Maximum size (TL)	IUCN Status	References
Dawkinsia arulius (Jerdon, 1849)	Freshwater	ENCR	Ornamental	12.0 cm	EN	Chandrashekhariah *et al.* (2000); Shahnawaz & Venkateshwarlu (2012); Shivashankar & Venkataramana (2012); Bhat & Hegde (2014); Venkateshwarlu *et al.* (2014)
Dawkinsia filamentosa (Valenciennes, 1844)	Freshwater; brackish	ENI	Ornamental	18.0 cm	LC	Chandrashekhariah *et al.* (2000); Shahnawaz & Venkateshwarlu (2012); Shivashankar & Venkataramana (2012); Naik *et al.* (2013b); Naik *et al.* (2013c); Bhat & Hegde (2014); Venkateshwarlu *et al.* (2014)
Esomus barbatus (Jerdon, 1849)	Freshwater	ENSI	Ornamental	12.0 cm	LC	Chandrashekhariah *et al.* (2000)
Esomus danricus (Hamilton, 1822)	Freshwater; brackish		Ornamental	13.0 cm	LC	Chandrashekhariah *et al.* (2000); Jen (2002); Thippeswamy *et al.* (2008)
Garra bicornuta (Narayan Rao, 1920)	Freshwater	ENKA, ENMH	Fishery doesn't exists	13.9 cm SL	NT	Chandrashekhariah *et al.* (2000); Shahnawaz & Venkateshwarlu (2012)
Garra gotyla gotyla (Gray, 1830)	Freshwater		Capture Fishery	18.0 cm	LC	Naik *et al.* (2013a); Naik *et al.* (2013b); Naik *et al.* (2013c); Naik *et al.* (2013d); Naik *et al.* (2014); Venkateshwarlu *et al.* (2014)

Order, Family, Scientific name	Environment	Endemic, Exotic	Human Use	Maximum size (TL)	IUCN Status	References
Garra gotyla stenorhynchus Jerdon, 1849	Freshwater	ENWG	Fishery doesn't exists	15.5 cm SL	LC	Chandrashekhariah *et al.* (2000); Jen (2002); Ramachandra *et al.* (2012); Bhat & Hegde (2014); Venkateshwarlu *et al.* (2014)
Garra hughi Silas, 1955	Freshwater	ENWG	Ornamental	15.5 cm SL	EN	Chandrashekhariah *et al.* (2000)
Garra kempi Hora, 1921	Freshwater		Fishery doesn't exists	11.3 cm SL	LC	Kumar (2013)
Garra lissorhynchus (McClelland, 1842)	Freshwater		Fishery doesn't exists	9.2 cm SL	LC	Venkateshwarlu *et al.* (2014)
Garra mcclellandi (Jerdon, 1849)	Freshwater	ENCR	Fishery doesn't exists	17.4 cm SL	LC	Chandrashekhariah *et al.* (2000); Jen (2002)
Garra mullya (Sykes, 1839)	Freshwater	ENI	Fishery doesn't exists	17.0 cm	LC	Chandrashekhariah *et al.* (2000); Shahnawaz & Venkateshwarlu (2012); Shivashankar & Venkataramana (2012); Bhat & Hegde (2014)
Haludaria fasciata (Jerdon, 1849)	Freshwater	ENSI	Ornamental	7.0 cm	LC	Chandrashekhariah *et al.* (2000); Jen (2002); Venkateshwarlu *et al.* (2014)
Hypophthalmichthys molitrix (Valenciennes, 1844)	Freshwater	EX	Aquaculture	105 cm	NT	Chandrashekhariah *et al.* (2000); Shivashankar & Venkataramana (2012); Naik *et al.* (2013a); Naik *et al.* (2013b); Naik *et al.* (2013c); Naik *et al.* (2013d); Naik *et al.* (2014)

Order, Family, Scientific name	Environment	Endemic, Exotic	Human Use	Maximum size (TL)	IUCN Status	References
Hypophthalmichthys nobilis (Richardson, 1845)	Freshwater	EX	Aquaculture	146 cm SL	DD	Chandrashekhariah *et al.* (2000)
Hypselobarbus curmuca (Hamilton, 1807)	Freshwater	ENI	Aquaculture	120 cm	EN	Chandrashekhariah *et al.* (2000); Jen (2002); Venkateshwarlu *et al.* (2014)
Hypselobarbus dobsoni (Day, 1876)	Freshwater	ENWG	Aquaculture	120 cm	DD	Jen (2002); Naik *et al.* (2013b); Naik *et al.* (2013c); Naik *et al.* (2013d)
Hypselobarbus dubius (Day, 1867)	Freshwater	ENI	Capture Fishery	25.0 cm	EN	Chandrashekhariah *et al.* (2000); Jen (2002); Ramachandra *et al.* (2012)
Hypselobarbus jerdoni (Day, 1870)	Freshwater	ENWG	Capture Fishery	46.0 cm	LC	Shahnawaz & Venkateshwarlu (2012); Bhat & Hegde (2014); Venkateshwarlu *et al.* (2014)
Hypselobarbus kolus (Sykes, 1839)	Freshwater	ENI	Fishery doesn't exists	30.0 cm	VU	Chandrashekhariah *et al.* (2000); Jen (2002); Vijaylaxmi *et al.* (2010); Vijaylaxmi & Vijaykumar (2011); Ramachandra *et al.* (2012); Shahnawaz & Venkateshwarlu (2012); Naik *et al.* (2013c); Venkateshwarlu *et al.* (2014)
Hypselobarbus kurali Menon & Rema Devi, 1995	Freshwater	ENI	Fishery doesn't exists	35.0 cm	LC	Venkateshwarlu *et al.* (2014)
Hypselobarbus lithopidos (Day, 1874)	Freshwater	ENI	Capture Fishery	60.0 cm	DD	Vijaylaxmi *et al.* (2010); Vijaylaxmi & Vijaykumar (2011); Shahnawaz & Venkateshwarlu (2012)
Hypselobarbus micropogon (Valenciennes, 1842)	Freshwater	ENCR	Gamefish	90.0 cm	EN	Chandrashekhariah *et al.* (2000); Jen (2002)

Order, Family, Scientific name	Environment	Endemic, Exotic	Human Use	Maximum size (TL)	IUCN Status	References
Hypselobarbus mussullah (Sykes, 1839)	Freshwater	ENI	Gamefish; Ornamental	150 cm	EN	Chandrashekhariah *et al.* (2000); Shahnawaz & Venkateshwarlu (2012)
Hypselobarbus pulchellus (Day, 1870)	Freshwater	ENKA	Fishery doesn't exists	44.0 cm	CR	Chandrashekhariah *et al.* (2000)
Labeo angra (Hamilton, 1822)	Freshwater		Capture Fishery	22.0 cm	LC	Shahnawaz & Venkateshwarlu (2012); Shivashankar & Venkataramana (2012); Naik *et al.* (2013c)
Labeo bata (Hamilton, 1822)	Freshwater		Aquaculture	61.0 cm	LC	Chandrashekhariah *et al.* (2000); Naik *et al.* (2013b); Naik *et al.* (2013c); Naik *et al.* (2014)
Labeo boga (Hamilton, 1822)	Freshwater		Ornamental	30.0 cm	LC	Chandrashekhariah *et al.* (2000)
Labeo boggut (Sykes, 1839)	Freshwater		Aquaculture	29.0 cm	LC	Chandrashekhariah *et al.* (2000); Vijaylaxmi & Vijaykumar (2011); Naik *et al.* (2013a)
Labeo calbasu (Hamilton, 1822)	Freshwater; brackish		Aquaculture	90.0 cm	LC	Chandrashekhariah *et al.* (2000); Shahnawaz & Venkateshwarlu (2012); Shivashankar & Venkataramana (2012); Naik *et al.* (2013b); Bhat & Hegde (2014); Venkateshwarlu *et al.* (2014)
Labeo dyocheilus (McClelland, 1839)	Freshwater		Capture Fishery	90.0 cm	LC	Naik *et al.* (2013c)

Order, Family, Scientific name	Environment	Endemic, Exotic	Human Use	Maximum size (TL)	IUCN Status	References
Labeo fimbriatus (Bloch, 1795)	Freshwater		Aquaculture	91.0 cm	LC	Chandrashekhariah *et al.* (2000); Vijaylaxmi & Vijaykumar (2011); Naik *et al.* (2013a); Naik *et al.* (2013b); Naik *et al.* (2013d); Bhat & Hegde (2014)
Labeo gonius (Hamilton, 1822)	Freshwater		Aquaculture	150 cm	LC	Naik *et al.* (2013b); Naik *et al.* (2013c)
Labeo kawrus (Sykes, 1839)	Freshwater	ENWG	Aquaculture	60.0 cm	LC	Thirumala *et al.* (2011)
Labeo kontius (Jerdon, 1849)	Freshwater	ENCR	Aquaculture	61.0 cm	LC	Chandrashekhariah *et al.* (2000); Jen (2002); Ramachandra *et al.* (2012); Naik *et al.* (2013c); Naik *et al.* (2013d)
Labeo pangusia (Hamilton, 1822)	Freshwater		Capture Fishery	90.0 cm	NT	Chandrashekhariah *et al.* (2000); Jen (2002)
Labeo porcellus (Heckel, 1844)	Freshwater		Capture Fishery	35.0 cm	LC	Chandrashekhariah *et al.* (2000); Jen (2002); Shahnawaz & Venkateshwarlu (2012)
Labeo potail (Sykes, 1839)	Freshwater	ENCR	Capture Fishery	30.0 cm	EN	Chandrashekhariah *et al.* (2000); Jen (2002); Shahnawaz & Venkateshwarlu (2012)
Labeo rohita (Hamilton, 1822)	Freshwater; brackish		Aquaculture; Gamefish	200 cm	LC	Chandrashekhariah *et al.* (2000); Vijaylaxmi & Vijaykumar (2011); Naik *et al.* (2013a); Naik *et al.* (2013b); Bhat & Hegde (2014)

Order, Family, Scientific name	Environment	Endemic, Exotic	Human Use	Maximum size (TL)	IUCN Status	References
Laubuka fasciata (Silas, 1958)	Freshwater	ENI	Fishery doesn't exists	6.0 cm	VU	Thippeswamy *et al.* (2008)
Laubuka laubuca (Hamilton, 1822)	Freshwater; brackish		Fishery doesn't exists	6.0 cm	VU	Chandrashekhariah *et al.* (2000); Thippeswamy *et al.* (2008)
Oreichthys cosuatis (Hamilton, 1822)	Freshwater		Fishery doesn't exists	8.0 cm	LC	Chandrashekhariah *et al.* (2000); Ramachandra *et al.* (2012); Naik *et al.* (2013c); Venkateshwarlu *et al.* (2014)
Osteobrama belangeri (Valenciennes, 1844)	Freshwater		Aquaculture	38.0 cm SL	NT	Chandrashekhariah *et al.* (2000)
Osteobrama cotio (Hamilton, 1822)	Freshwater		Fishery doesn't exists	15.0 cm	LC	Chandrashekhariah *et al.* (2000); Vijaylaxmi *et al.* (2010); Vijaylaxmi & Vijaykumar (2011); Shahnawaz & Venkateshwarlu (2012); Naik *et al.* (2013a); Naik *et al.* (2013c); Naik *et al.* (2013d)
Osteobrama cunma (Day, 1888)	Freshwater		Fishery doesn't exists	15.0 cm	LC	Chandrashekhariah *et al.* (2000); Thirumala *et al.* (2011); Kumar (2013)
Osteobrama neilli (Day, 1873)	Freshwater	ENI	Fishery doesn't exists	12.0 cm	LC	Chandrashekhariah *et al.* (2000); Thirumala *et al.* (2011); Shahnawaz & Venkateshwarlu (2012)
Osteobrama peninsularis Silas, 1952	Freshwater	ENPI	Capture Fishery	15.0 cm	DD	Chandrashekhariah *et al.* (2000)

Order, Family, Scientific name	Environment	Endemic, Exotic	Human Use	Maximum size (TL)	IUCN Status	References
Osteobrama vigorsii (Sykes, 1839)	Freshwater; brackish	ENI	Capture Fishery	30.0 cm	LC	Chandrashekhariah *et al.* (2000)
Osteochilichthys brevidorsalis (Day, 1873)	Freshwater	ENI	Fishery doesn't exists	15.0 cm	LC	Chandrashekhariah *et al.* (2000); Jen (2002)
Osteochilichthys thomassi (Day, 1877)	Freshwater	ENWG, ENKA	Aquaculture	32.0 cm	LC	Chandrashekhariah *et al.* (2000); Jen (2002); Shahnawaz & Venkateshwarlu (2012)
Osteochilus nashii (Day, 1869)	Freshwater	ENI	Capture Fishery	18.0 cm	LC	Chandrashekhariah *et al.* (2000); Ramachandra *et al.* (2012); Shahnawaz & Venkateshwarlu (2012); Shivashankar & Venkataramana (2012)
Parapsilorhynchus prateri Hora & Misra, 1938	Freshwater	ENI	Fishery doesn't exists	11.0 cm	CR	Chandrashekhariah *et al.* (2000)
Pethia conchonius (Hamilton, 1822)	Freshwater		Ornamental	14.0 cm	LC	Chandrashekhariah *et al.* (2000); Jen (2002); Shahnawaz & Venkateshwarlu (2012); Naik *et al.* (2013c)
Pethia guganio (Hamilton, 1822)	Freshwater		Fishery doesn't exists	8.0 cm	LC	Chandrashekhariah *et al.* (2000)
Pethia narayani (Hora, 1937))	Freshwater	ENI	Fishery doesn't exists	7.8 cm	LC	Chandrashekhariah *et al.* (2000); Jen (2002); Anandhi & Sharath (2014); Bhat & Hegde (2014); Venkateshwarlu *et al.* (2014)

Order, Family, Scientific name	Environment	Endemic, Exotic	Human Use	Maximum size (TL)	IUCN Status	References
Pethia ticto (Hamilton, 1822)	Freshwater; brackish		Ornamental	10.0 cm	LC	Chandrashekhariah *et al.* (2000); Vijaylaxmi & Vijaykumar (2011); Shahnawaz & Venkateshwarlu (2012); Naik *et al.* (2013b); Naik *et al.* (2013c); Naik *et al.* (2013d);
Puntius amphibius (Valenciennes, 1842)	Freshwater; brackish		Capture Fishery	20.0 cm	DD	Chandrashekhariah *et al.* (2000); Bhat & Hegde (2014); Shahnawaz & Venkateshwarlu (2012); Shivashankar & Venkataramana (2012)
Puntius bimaculatus (Bleeker, 1863)	Freshwater		Ornamental	7.0 cm	LC	Chandrashekhariah *et al.* (2000)
Puntius cauveriensis Hora, 1937	Freshwater	ENKA	Fishery doesn't exists	7.4 cm	EN	Chandrashekhariah *et al.* (2000); Jen (2002)
Puntius chelynoides (McClelland, 1839)	Freshwater	ENI	Fishery doesn't exists	66.0 cm	VU	Vijaylaxmi *et al.* (2010); Vijaylaxmi & Vijaykumar (2011); Naik *et al.* (2013a)
Puntius chola (Hamilton, 1822)	Freshwater		Ornamental	15.0 cm	LC	Chandrashekhariah *et al.* (2000); Shahnawaz & Venkateshwarlu (2012); Shivashankar & Venkataramana (2012); Kumar (2013); Naik *et al.* (2013c); Naik *et al.* (2013d); Bhat & Hegde (2014)
Puntius deccanensis Yazdani & Babu Rao, 1976	Freshwater	ENI	Fishery doesn't exists	3.7 cm SL	CR	Thirumala *et al.* (2011)

Order, Family, Scientific name	Environment	Endemic, Exotic	Human Use	Maximum size (TL)	IUCN Status	References
Puntius dorsalis (Jerdon, 1849)	Freshwater		Ornamental	25.0 cm	LC	Chandrashekhariah *et al.* (2000); Jen (2002); Ramachandra *et al.* (2012); Venkateshwarlu *et al.* (2014)
Puntius melanostigma (Day, 1878)	Freshwater	ENI	Fishery doesn't exists	10.0 cm	NE	Chandrashekhariah *et al.* (2000); Jen (2002)
Puntius parrah Day, 1865	Freshwater	ENWG	Capture Fishery	15.0 cm	LC	Chandrashekhariah *et al.* (2000); Jen (2002); Venkateshwarlu *et al.* (2014)
Puntius sahyadriensis Silas, 1953	Freshwater	ENWG	Ornamental	7.0 cm	LC	Chandrashekhariah *et al.* (2000); Shahnawaz & Venkateshwarlu (2012); Venkateshwarlu *et al.* (2014)
Puntius sophore (Hamilton, 1822)	Freshwater; brackish		Ornamental	20.0 cm	LC	Chandrashekhariah *et al.* (2000); Thirumala *et al.* (2011); Shahnawaz & Venkateshwarlu (2012); Kumar (2013); Naik *et al.* (2013c)
Puntius thomassi (Day, 1874)	Freshwater	ENKA, ENKL	Capture Fishery	100.0 cm	LC	Chandrashekhariah *et al.* (2000); Shahnawaz & Venkateshwarlu (2012)
Puntius vittatus Day, 1865	Freshwater; brackish		Ornamental	5.0 cm	LC	Chandrashekhariah *et al.* (2000); Naik *et al.* (2013b); Venkateshwarlu *et al.* (2014)
Raiamas bola (Hamilton, 1822)	Freshwater		Gamefish	35.0 cm	LC	Naik *et al.* (2013a); Naik *et al.* (2013c); Naik *et al.* (2014)

Order, Family, Scientific name	Environment	Endemic, Exotic	Human Use	Maximum size (TL)	IUCN Status	References
Rasbora caverii (Jerdon, 1849)	Freshwater; brackish		Ornamental	10.0 cm	LC	Chandrashekhariah *et al.* (2000); Jen (2002); Venkateshwarlu *et al.* (2014)
Rasbora daniconius (Hamilton, 1822)	Freshwater; brackish		Ornamental	15.0 cm	LC	Chandrashekhariah *et al.* (2000); Thippeswamy *et al.* (2008); Vijaylaxmi *et al.* (2010); Vijaylaxmi & Vijaykumar (2011); Shahnawaz & Venkateshwarlu (2012); Shivashankar & Venkataramana (2012); Kumar (2013); Naik *et al.* (2013a); Naik *et al.* (2013b); Naik *et al.* (2013c); Venkateshwarlu *et al.* (2014)
Rasbora dusonensis (Bleeker, 1850)	Freshwater	EX	Ornamental	12.0 cm SL	NE	Venkateshwarlu *et al.* (2014)
Rasbora labiosa Mukerji, 1935	Freshwater	ENI	Ornamental	8.5 cm	LC	Chandrashekhariah *et al.* (2000)
Rasbora rasbora (Hamilton, 1822)	Freshwater; brackish		Ornamental	13.0 cm	LC	Chandrashekhariah *et al.* (2000); Bhat & Hegde (2014); Venkateshwarlu *et al.* (2014)
Rohtee ogilbii Sykes, 1839	Freshwater	ENI	Capture Fishery	50.0 cm	LC	Chandrashekhariah *et al.* (2000); Shahnawaz & Venkateshwarlu (2012)
Salmophasia acinaces (Valenciennes, 1844)	Freshwater	ENI	Fishery doesn't exists	15.0 cm	LC	Chandrashekhariah *et al.* (2000); Jen (2002); Venkateshwarlu *et al.* (2014)

Order, Family, Scientific name	Environment	Endemic, Exotic	Human Use	Maximum size (TL)	IUCN Status	References
Salmophasia bacaila (Hamilton, 1822)	Freshwater; brackish		Fishery doesn't exists	18.0 cm	LC	Vijaylaxmi *et al.* (2010); Vijaylaxmi & Vijaykumar (2011); Naik *et al.* (2013a); Naik *et al.* (2013c); Naik *et al.* (2013d); Naik *et al.* (2014)
Salmophasia balookee (Sykes, 1839)	Freshwater		Capture Fishery	15.0 cm	LC	Chandrashekhariah *et al.* (2000); Jen (2002)
Salmophasia belachi (Jayaraj, Krishna Rao, Ravichandra Reddy, Shakuntala & Devaraj, 1999)	Freshwater	ENKA	Capture Fishery	12.0 cm	VU	Jen (2002)
Salmophasia boopis (Day, 1874)	Freshwater	ENI	Fishery doesn't exists	12.0 cm	LC	Chandrashekhariah *et al.* (2000); Ramachandra *et al.* (2012); Shahnawaz & Venkateshwarlu (2012); Venkateshwarlu *et al.* (2014)
Salmophasia horai (Silas, 1951)	Freshwater	ENI	Fishery doesn't exists	10.0 cm	VU	Chandrashekhariah *et al.* (2000); Jen (2002)
Salmophasia novacula (Valenciennes, 1840)	Freshwater	ENI	Fishery doesn't exists	12.5 cm	LC	Chandrashekhariah *et al.* (2000); Jen (2002)
Salmophasia phulo (Hamilton, 1822)	Freshwater		Fishery doesn't exists	12.0 cm	LC	Chandrashekhariah *et al.* (2000); Jen (2002)
Salmophasia sardinella (Valenciennes, 1844)	Freshwater		Fishery doesn't exists	15.0 cm	LC	Shahnawaz & Venkateshwarlu (2012)
Salmophasia untrahi (Day, 1869)	Freshwater	ENI	Fishery doesn't exists	20.0 cm	LC	Chandrashekhariah *et al.* (2000); Jen (2002); Thirumala *et al.* (2011); Shivashankar & Venkataramana (2012); Kumar (2013)

Order, Family, Scientific name	Environment	Endemic, Exotic	Human Use	Maximum size (TL)	IUCN Status	References
Schismatorhynchos nukta (Sykes, 1839)	Freshwater	ENKA, ENMH	Capture Fishery	30.0 cm	EN	Chandrashekhariah *et al.* (2000)
Systomus sarana (Hamilton, 1822)	Freshwater; brackish		Gamefish; Ornamental	42.0 cm	LC	Chandrashekhariah *et al.* (2000); Jen (2002); Shahnawaz & Venkateshwarlu (2012); Ramachandra *et al.* (2012); Venkateshwarlu *et al.* (2014)
Systomus spilurus (Günther, 1868)	Freshwater		Fishery doesn't exists	–	NE	Venkateshwarlu *et al.* (2014)
Thynnichthys sandkhol (Sykes, 1839)	Freshwater	ENPI	Aquaculture	46.0 cm	EN	Chandrashekhariah *et al.* (2000)
Tor khudree (Sykes, 1839)	Freshwater		Aquaculture; Gamefish	50.0 cm	EN	Chandrashekhariah *et al.* (2000)
Tor tor (Hamilton, 1822)	Freshwater		Aquaculture; Gamefish	200 cm	NT	Bhat & Hegde (2014)
Family: Nemacheilidae						
Acanthocobitis botia (Hamilton, 1822)	Freshwater		Fishery doesn't exists	11.0 cm	LC	Ramachandra *et al.* (2012); Shivashankar & Venkataramana (2012)
Acanthocobitis mooreh (Sykes, 1839)	Freshwater	ENPI	Fishery doesn't exists	4.4 cm SL	LC	Chandrashekhariah *et al.* (2000)
Acanthocobitis rubidipinnis (Blyth, 1860)	Freshwater		Fishery doesn't exists	10.0 cm	LC	Chandrashekhariah *et al.* (2000)
Indoreonectes evezardi (Day, 1872)	Freshwater	ENWG	Fishery doesn't exists	3.8 cm SL	LC	Chandrashekhariah *et al.* (2000)

Order, Family, Scientific name	Environment	Endemic, Exotic	Human Use	Maximum size (TL)	IUCN Status	References
Longischistura striata (Day, 1867)	Freshwater	ENKA, ENKL	Fishery doesn't exists	5.0 cm SL	LC	Chandrashekhariah *et al.* (2000); Ramachandra *et al.* (2012); Shivashankar & Venkataramana Jen (2002); Thirumala *et al.* (2011)
Nemacheilus anguilla Annandale, 1919	Freshwater	ENWG	Fishery doesn't exists	5.0 cm SL	LC	Chandrashekhariah *et al.* (2000); Ramachandra *et al.* (2012); Shivashankar & Venkataramana (2012)
Nemacheilus denisoni Day, 1867	Freshwater	ENI	Fishery doesn't exists	5.0 cm SL	LC	Chandrashekhariah *et al.* (2000); Ramachandra *et al.* (2012); Shivashankar & Venkataramana (2012); Venkateshwarlu *et al.* (2014)
Nemacheilus guentheri Day, 1867	Freshwater	ENPI	Ornamental	5.6 cm SL	LC	Chandrashekhariah *et al.* (2000); Jen (2002); Bhat & Hegde (2014)
Nemacheilus kodaguensis Menon, 1987	Freshwater	ENCR	Fishery doesn't exists	3.6 cm SL	LC	Chandrashekhariah *et al.* (2000); Jen (2002)
Nemacheilus nilgiriensis Menon, 1987	Freshwater	ENWG, ENTN	Fishery doesn't exists	5.1 cm SL	LC	Ramachandra *et al.* (2012); Venkateshwarlu *et al.* (2014)
Nemacheilus pulchellus Day, 1873	Freshwater	ENWG	Fishery doesn't exists	4.6 cm SL	EN	Chandrashekhariah *et al.* (2000); Jen (2002)
Nemacheilus rueppelli (Sykes, 1839)	Freshwater	ENWG, ENKA, ENMH	Fishery doesn't exists	7.4 cm SL	LC	Chandrashekhariah *et al.* (2000); Shahnawaz & Venkateshwarlu (2012); Naik *et al.* (2013a); Naik *et al.* (2013b); Naik *et al.* (2013c); Naik *et al.* (2013d); Naik *et al.* (2014)

Order, Family, Scientific name	Environment	Endemic, Exotic	Human Use	Maximum size (TL)	IUCN Status	References
Nemacheilus semiarmatus Day, 1867	Freshwater	ENPI	Aquaculture; Ornamental	5.6 cm SL	LC	Chandrashekhariah *et al.* (2000); Jen (2002); Ramachandra *et al.* (2012); Shahnawaz & Venkateshwarlu (2012); Shivashankar & Venkataramana (2012)
Nemacheilus stigmofasciatus Arunachalam & Muralidharan, 2009	Freshwater	ENWG	Fishery doesn't exists	3.8 cm SL	NE	Venkateshwarlu *et al.* (2014)
Schistura nagodiensis Sreekantha, Gururaja, Rema Devi, Indra & Ramachandra, 2006	Freshwater	ENWG	Fishery doesn't exists	2.8 cm SL	EN	Venkateshwarlu *et al.* (2014)
Family: Psilorhynchidae						
Psilorhynchus tenure Arunachalam & Muralidharan, 2008	Freshwater	ENKA	Fishery doesn't exists	6.31 cm	NE	Shahnawaz & Venkateshwarlu (2012)
Order: Cyprinodontiformes						
Family: Aplocheilidae						
Aplocheilus lineatus (Valenciennes, 1846)	Freshwater; brackish	ENPI	Ornamental	10.0 cm	LC	Chandrashekhariah *et al.* (2000); Ramachandra *et al.* (2012); Bhat & Hegde (2014); Venkateshwarlu *et al.* (2014)
Family: Poeciliidae						
Gambusia affinis (Baird & Girard, 1853)	Freshwater; brackish	EX	Ornamental	4.0 cm	LC	Chandrashekhariah *et al.* (2000); Thirumala *et al.* (2011); Kumar (2013)

Order, Family, Scientific name	Environment	Endemic, Exotic	Human Use	Maximum size (TL)	IUCN Status	References
Order: Elopiformes						
Family: Megalopidae						
Megalops cyprinoides (Broussonet, 1782)	Marine; freshwater; brackish		Aquaculture; Gamefish	45.5 cm SL	DD	Chandrashekhariah *et al.* (2000); Jen (2002); Venkateshwarlu *et al.* (2014);
Order: Gonorynchiformes						
Family: Chanidae						
Chanos chanos (Forsskal, 1775)	Marine; freshwater; brackish		Aquaculture; Gamefish	180 cm SL	NE	Chandrashekhariah *et al.* (2000); Jen (2002)
Order: Mugiliformes						
Family: Mugilidae						
Rhinomugil corsula (Hamilton, 1822)	Freshwater; brackish		Aquaculture	45.0 cm	LC	Chandrashekhariah *et al.* (2000)
Order: Osteoglossiformes						
Family: Notopteridae						
Chitala chitala (Hamilton, 1822)	Freshwater		Aquaculture; Gamefish; Ornamental	122 cm SL	NT	Naik *et al.* (2013c); Naik *et al.* (2013c); Naik *et al.* (2014)
Notopterus notopterus (Pallas, 1769)	Freshwater; brackish		Aquaculture; Ornamental	60.0 cm SL	LC	Chandrashekhariah *et al.* (2000); Vijaylaxmi & Vijaykumar (2011); Shivashankar & Venkataramana (2012); Naik *et al.* (2013a)

Order, Family, Scientific name	Environment	Endemic, Exotic	Human Use	Maximum size (TL)	IUCN Status	References
Order: Perciformes						
Family: Ambassidae						
Ambassis kopsii Bleeker, 1858	Marine; freshwater; brackish		Capture Fishery	10.2 cm	NE	Thirumala *et al.* (2011); Kumar (2013)
Chanda nama Hamilton, 1822	Freshwater; brackish		Ornamental	11.0 cm	LC	Chandrashekhariah *et al.* (2000); Shivashankar & Venkataramana (2012); Naik *et al.* (2013a); Naik *et al.* (2013c); Naik *et al.* (2013d); Naik *et al.* (2014); Venkateshwarlu *et al.* (2014)
Parambassis baculis (Hamilton, 1822)	Freshwater	ENI	Fishery doesn't exists	5.0 cm SL	LC	Naik *et al.* (2013c); Naik *et al.* (2013d)
Parambassis ranga (Hamilton, 1822)	Freshwater; brackish		Ornamental	8.0 cm	LC	Chandrashekhariah *et al.* (2000); Ramachandra *et al.* (2012); Naik *et al.* (2013c); Naik *et al.* (2013d); Bhat & Hegde (2014); Naik *et al.* (2014)
Parambassis thomassi (Day, 1870)	Marine; freshwater; brackish	ENWG, ENKA, ENKL	Capture Fishery	12.0 cm SL	LC	Venkateshwarlu *et al.* (2014)
Family: Anabantidae						
Anabas testudineus (Bloch, 1792)	Freshwater; brackish		Aquaculture; Ornamental	25.0 cm	DD	Naik *et al.* (2013c); Naik *et al.* (2013d)

Order, Family, Scientific name	Environment	Endemic, Exotic	Human Use	Maximum size (TL)	IUCN Status	References
Family: Badidae						
Dario huli Britz & Ali, 2015	Freshwater	ENKA	Fishery doesn't exists	-	NE	Britz & Ali (2015)
Family: Channidae						
Channa gachua (Hamilton, 1822)	Freshwater		Ornamental	20.0 cm SL	LC	Naik *et al.* (2013b); Naik *et al.* (2013c)
Channa marulius (Hamilton, 1822)	Freshwater		Aquaculture; Gamefish; Ornamental	183 cm	LC	Chandrashekhariah *et al.* (2000); Vijaylaxmi & Vijaykumar (2011); Naik *et al.* (2013a); Naik *et al.* (2013c); Naik *et al.* (2014); Venkateshwarlu *et al.* (2014)
Channa orientalis Bloch & Schneider, 1801	Freshwater; brackish		Ornamental	33.0 cm	NE	Chandrashekhariah *et al.* (2000); Ramachandra *et al.* (2012); Shivashankar & Venkataramana (2012); Venkateshwarlu *et al.* (2014)
Channa punctata (Bloch, 1793)	Freshwater; brackish		Aquaculture; Ornamental	31.0 cm	LC	Chandrashekhariah *et al.* (2000); Vijaylaxmi & Vijaykumar (2011); Kumar (2013); Naik *et al.* (2013c); Naik *et al.* (2014);
Channa striata (Bloch, 1793)	Freshwater; brackish		Aquaculture; Ornamental	100.0 cm SL	LC	Chandrashekhariah *et al.* (2000); Vijaylaxmi & Vijaykumar (2011); Shivashankar & Venkataramana (2012); Naik *et al.* (2013a); Naik *et al.* (2013b); Naik *et al.* (2013d)

Order, Family, Scientific name	Environment	Endemic, Exotic	Human Use	Maximum size (TL)	IUCN Status	References
Family: Cichlidae						
Etroplus canarensis Day, 1877	Freshwater	ENKA	Fishery doesn't exists	11.5 cm	EN	Chandrashekhariah *et al.* (2000); Anandhi & Sharath (2014); Venkateshwarlu *et al.* (2014)
Etroplus maculatus (Bloch, 1795)	Freshwater; brackish		Ornamental	8.0 cm	LC	Chandrashekhariah *et al.* (2000); Venkateshwarlu *et al.* (2014)
Etroplus suratensis (Bloch, 1790)	Freshwater; brackish		Aquaculture; Ornamental	40.0 cm	LC	Chandrashekhariah *et al.* (2000); Jen (2002); Abida *et al.* (2009); Venkateshwarlu *et al.* (2014)
Oreochromis mossambicus (Peters, 1852)	Freshwater; brackish	EX	Aquaculture; Gamefish; Ornamental	39.0 cm SL	NT	Chandrashekhariah *et al.* (2000); Vijaylaxmi & Vijaykumar (2011); Ramachandra *et al.* (2012); Shivashankar & Venkataramana (2012); Kumar (2013); Naik *et al.* (2013b); Naik *et al.* (2013c); Naik *et al.* (2014)
Oreochromis niloticus (Linnaeus, 1758)	Freshwater; brackish	EX	Aquaculture	60.0 cm SL	NE	Naik *et al.* (2013b); Naik *et al.* (2013c); Naik *et al.* (2013d); Naik *et al.* (2014)
Family: Gerreidae						
Gerres erythrourus (Bloch, 1791)	Marine; freshwater; brackish		Capture Fishery	30.0 cm	NE	Venkateshwarlu *et al.* (2014)

Order, Family, Scientific name	Environment	Endemic, Exotic	Human Use	Maximum size (TL)	IUCN Status	References
Gerres filamentosus Cuvier, 1829	Marine; freshwater; brackish		Capture Fishery	35.0 cm	LC	Venkateshwarlu *et al.* (2014)
Gerres limbatus Cuvier, 1830	Marine; freshwater; brackish		Capture Fishery	15.0 cm	LC	Venkateshwarlu *et al.* (2014)
Family: Gobidae						
Glossogobius giuris (Hamilton, 1822)	Marine; freshwater; brackish		Aquaculture; Ornamental	50.0 cm SL	LC	Chandrashekhariah *et al.* (2000); Ramachandra *et al.* (2012); Shivashankar & Venkataramana (2012); Naik *et al.* (2013d); Naik *et al.* (2014); Venkateshwarlu *et al.* (2014)
Psammogobius biocellatus (Valenciennes, 1837)	Marine; freshwater; brackish		Fishery doesn't exists	12.0 cm	LC	Vijaylaxmi *et al.* (2010); Vijaylaxmi & Vijaykumar (2011); Venkateshwarlu *et al.* (2014)
Family: Nandidae						
Nandus nandus (Hamilton, 1822)	Freshwater; brackish		Ornamental	20.0 cm	LC	Abida *et al.* (2009)
Family: Osphronemidae						
Macropodus spechti Schreitmüller, 1936	Freshwater	EX	Fishery doesn't exists	5.8 cm SL	DD	Venkateshwarlu *et al.* (2014)
Osphronemus goramy Lacepède, 1801	Freshwater; brackish	EX	Aquaculture; Ornamental	70.0 cm SL	LC	Chandrashekhariah *et al.* (2000); Thippeswamy *et al.* (2008)

Order, Family, Scientific name	Environment	Endemic, Exotic	Human Use	Maximum size (TL)	IUCN Status	References
Pseudosphromenus cupanus (Cuvier, 1831)	Freshwater; brackish		Ornamental	7.5 cm	LC	Chandrashekhariah *et al.* (2000); Ramachandra *et al.* (2012); Venkateshwarlu *et al.* (2014)
Family: Pristolepididae						
Pristolepis marginata Jerdon, 1849	Freshwater	ENWG, ENKL	Ornamental	15.0 cm SL	LC	Chandrashekhariah *et al.* (2000); Venkateshwarlu *et al.* (2014)
Order: Siluriformes						
Family: Amblycipitidae						
Amblyceps mangois (Hamilton, 1822)	Freshwater		Fishery doesn't exists	12.5 cm SL	LC	Chandrashekhariah *et al.* (2000)
Family: Bagridae						
Batasio sharavatiensis Bhatt & Jayaram, 2004	Freshwater	ENSR-	Fishery doesn't exists	10.7 cm SL	EN	Ramachandra *et al.* (2012)
Hemibagrus maydelli (Rössel, 1964)	Freshwater	ENKR	Fishery doesn't exists	165 cm	LC	Chandrashekhariah *et al.* (2000)
Hemibagrus menoda (Hamilton, 1822)	Freshwater	ENI	Fishery doesn't exists	45.0 cm	LC	Naik *et al.* (2013c)
Hemibagrus punctatus (Jerdon, 1849)	Freshwater	ENCR	Capture Fishery	45.0 cm	CR	Chandrashekhariah *et al.* (2000); Jen (2002)
Horabagrus brachysoma (Günther, 1864)	Freshwater; brackish	ENWG, ENKL	Ornamental	45.0 cm	VU	Venkateshwarlu *et al.* (2014)

Order, Family, Scientific name	Environment	Endemic, Exotic	Human Use	Maximum size (TL)	IUCN Status	References
Mystus armatus (Day, 1865)	Freshwater; brackish		Ornamental	14.5 cm SL	LC	Shivashankar & Venkataramana (2012); Venkateshwarlu *et al.* (2014)
Mystus bleekeri (Day, 1877)	Freshwater		Ornamental	15.5 cm	LC	Vijaylaxmi *et al.* (2010); Vijaylaxmi & Vijaykumar (2011); Ramachandra *et al.* (2012); Naik *et al.* (2013a); Naik *et al.* (2013c); Naik *et al.* (2013d)
Mystus cavasius (Hamilton, 1822)	Freshwater; brackish		Capture Fishery	40.0 cm SL	LC	Chandrashekhariah *et al.* (2000); Shivashankar & Venkataramana (2012); Kumar (2013); Naik *et al.* (2013b); Naik *et al.* (2013c); Naik *et al.* (2013d); Venkateshwarlu *et al.* (2014)
Mystus gulio (Hamilton, 1822)	Freshwater; brackish		Capture Fishery	46.0 cm	LC	Chandrashekhariah *et al.* (2000); Vijaylaxmi *et al.* (2010); Vijaylaxmi & Vijaykumar (2011); Venkateshwarlu *et al.* (2014)
Mystus keletius (Valenciennes, 1840)	Freshwater		Ornamental	18.0 cm	LC	Ramachandra *et al.* (2012)
Mystus malabaricus (Jerdon, 1849)	Freshwater; brackish	ENWG	Capture Fishery	15.0 cm	NT	Chandrashekhariah *et al.* (2000); Jen (2002); Ramachandra *et al.* (2012); Venkateshwarlu *et al.* (2014)
Mystus montanus (Jerdon, 1849)	Freshwater; brackish	ENPI	Capture Fishery	15.0 cm	LC	Chandrashekhariah *et al.* (2000); Venkateshwarlu *et al.* (2014)

Order, Family, Scientific name	Environment	Endemic, Exotic	Human Use	Maximum size (TL)	IUCN Status	References
Mystus oculatus (Valenciennes, 1840)	Freshwater; brackish	ENKL, ENTN	Capture Fishery	15.0 cm	LC	Venkateshwarlu *et al.* (2014)
Mystus tengara (Hamilton, 1822)	Freshwater		Fishery doesn't exists	18.0 cm	LC	Naik *et al.* (2013c); Naik *et al.* (2013d); Naik *et al.* (2014)
Mystus vittatus (Bloch, 1794)	Freshwater; brackish		Ornamental	21.0 cm SL	LC	Chandrashekhariah *et al.* (2000); Jen (2002); Shivashankar & Venkataramana (2012); Naik *et al.* (2013c); Naik *et al.* (2013d); Naik *et al.* (2014)
Rita gogra (Sykes, 1839)	Freshwater	ENKR	Capture Fishery	26.0 cm	LC	Chandrashekhariah *et al.* (2000); Shivashankar & Venkataramana (2012); Naik *et al.* (2013d); Naik *et al.* (2013b); Naik *et al.* (2013c); Naik *et al.* (2014)
Rita kuturnee (Sykes, 1839)	Freshwater	ENPI, ENKR	Capture Fishery	30.0 cm	LC	Chandrashekhariah *et al.* (2000)
Rita rita (Hamilton, 1822)	Freshwater; brackish		Capture Fishery	150 cm	LC	Vijaylaxmi *et al.* (2010); Vijaylaxmi & Vijaykumar (2011); Naik *et al.* (2013a); Naik *et al.* (2013c); Naik *et al.* (2013d)
Sperata aor (Hamilton, 1822)	Freshwater		Gamefish	180 cm	LC	Chandrashekhariah *et al.* (2000); Naik *et al.* (2013b); Naik *et al.* (2013c); Naik *et al.* (2013d); Naik *et al.* (2014)

Order, Family, Scientific name	Environment	Endemic, Exotic	Human Use	Maximum size (TL)	IUCN Status	References
Sperata seenghala (Sykes, 1839)	Freshwater; brackish		Aquaculture; Gamefish	150 cm	LC	Chandrashekhariah *et al.* (2000); Vijaylaxmi *et al.* (2010); Vijaylaxmi & Vijaykumar (2011); Naik *et al.* (2013a); Naik *et al.* (2013b); Naik *et al.* (2013c); Naik *et al.* (2013d); Naik *et al.* (2014)
Family: Claridae						
Clarias batrachus (Linnaeus, 1758)	Freshwater; brackish		Aquaculture; Ornamental	47.0 cm	LC	Shivashankar & Venkataramana (2012); Kumar (2013); Naik *et al.* (2013a); Naik *et al.* (2013b); Naik *et al.* (2014)
Clarias dussumieri Valenciennes, 1840	Freshwater	ENSI	Capture Fishery	25.0 cm	NT	Chandrashekhariah *et al.* (2000); Ramachandra *et al.* (2012); Shivashankar & Venkataramana (2012)
Clarias gariepinus (Burchell, 1822)	Freshwater	EX	Aquaculture; Gamefish	170 cm	LC	Naik *et al.* (2013b); Naik *et al.* (2013c); Naik *et al.* (2013d); Naik *et al.* (2014)
Family: Heteropneustidae						
Heteropneustes fossilis (Bloch, 1794)	Freshwater; brackish		Aquaculture; Ornamental	30.0 cm	LC	Chandrashekhariah *et al.* (2000); Shivashankar & Venkataramana (2012); Naik *et al.* (2013a); Naik *et al.* (2013d); Naik *et al.* (2014)

Order, Family, Scientific name	Environment	Endemic, Exotic	Human Use	Maximum size (TL)	IUCN Status	References
Family: Loricardiidae						
Hypostomus plecostomus (Linnaeus, 1758)	Freshwater	EX	Ornamental	50.0 cm SL	NE	Naik *et al.* (2013b)
Family: Pangasiidae						
Pangasius pangasius (Hamilton, 1822)	Freshwater; brackish		Aquaculture; Gamefish	300 cm SL	LC	Chandrashekhariah *et al.* (2000); Naik *et al.* (2013a); Naik *et al.* (2013c); Naik *et al.* (2013d); Naik *et al.* (2014)
Family: Schilbeidae						
Ailia coila (Hamilton, 1822)	Freshwater; brackish		Capture Fishery	30.0 cm	NT	Naik *et al.* (2013c); Naik *et al.* (2013d); Naik *et al.* (2014)
Clupisoma garua (Hamilton, 1822)	Freshwater; brackish		Gamefish	60.9 cm SL	LC	Naik *et al.* (2013c)
Eutropiichthys goongwaree (Sykes, 1839)	Freshwater	ENPI	Fishery doesn't exists	64.0 cm	DD	Chandrashekhariah *et al.* (2000)
Eutropiichthys vacha (Hamilton, 1822)	Freshwater; brackish		Gamefish	34.0 cm	LC	Chandrashekhariah *et al.* (2000); Naik *et al.* (2013c)
Neotropius atherinoides (Bloch, 1794)	Freshwater; brackish		Ornamental	15.0 cm	LC	Ramachandra *et al.* (2012)
Neotropius khavalchor Kulkarni, 1952	Freshwater	ENKR	Fishery doesn't exists	15.0 cm	DD	Chandrashekhariah *et al.* (2000)
Proeutropiichthys taakree (Sykes, 1839)	Freshwater	ENWG, ENKL, ENMH	Capture Fishery	42.0 cm	LC	Chandrashekhariah *et al.* (2000)
Pseudeutropius mitchelli Günther, 1864	Freshwater	ENI	Ornamental	90.0 cm	EN	Venkateshwarlu *et al.* (2014)

Order, Family, Scientific name	Environment	Endemic, Exotic	Human Use	Maximum size (TL)	IUCN Status	References
Silonia childreni (Sykes, 1839)	Freshwater	ENWG	Capture Fishery	48.0 cm	EN	Chandrashekhariah *et al.* (2000); Jen (2002)
Silonia silondia (Hamilton, 1822)	Freshwater; brackish		Gamefish	183 cm	LC	Naik *et al.* (2013a); Naik *et al.* (2013b); Naik *et al.* (2013d); Naik *et al.* (2014)
Family: Siluridae						
Ompok bimaculatus (Bloch, 1794)	Freshwater; brackish		Aquaculture; Ornamental	45.0 cm SL	NT	Chandrashekhariah *et al.* (2000); Vijaylaxmi & Vijaykumar (2011); Kumar (2013); Naik *et al.* (2013a); Naik *et al.* (2013b); Naik *et al.* (2013d); Naik *et al.* (2014)
Ompok malabaricus (Valenciennes, 1840)	Freshwater	ENWG	Aquaculture	51.0 cm SL	LC	Chandrashekhariah *et al.* (2000); Venkateshwarlu *et al.* (2014)
Ompok pabda (Hamilton, 1822)	Freshwater		Capture Fishery	30.0 cm	NT	Chandrashekhariah *et al.* (2000); Vijaylaxmi *et al.* (2010); Vijaylaxmi & Vijaykumar (2011); Naik *et al.* (2013a); Naik *et al.* (2013b); Naik *et al.* (2013c); Naik *et al.* (2013d); Naik *et al.* (2014); Venkateshwarlu *et al.* (2014)
Ompok pabo (Hamilton, 1822)	Freshwater		Capture Fishery	25.0 cm	NT	Ramachandra *et al.* (2012); Venkateshwarlu *et al.* (2014)
Pterocryptis wynaadensis (Day, 1873)	Freshwater	ENWG, ENKA, ENKL	Fishery doesn't exists	30.0 cm	EN	Jen (2002); Shivashankar & Venkataramana (2012)

Order, Family, Scientific name	Environment	Endemic, Exotic	Human Use	Maximum size (TL)	IUCN Status	References
Wallago attu (Bloach&Schneider,1801)	Freshwater; brackish		Gamefish	240 cm	NT	Chandrashekhariah *et al.* (2000); Shivashankar & Venkataramana (2012); Naik *et al.* (2013b); Naik *et al.* (2013c); Naik *et al.* (2014)
Family: Sisoridae						
Bagarius bagarius (Hamilton, 1822)	Freshwater; brackish		Gamefish	200 cm	NT	Naik *et al.* (2013c); Naik *et al.* (2013d); Naik *et al.* (2014)
Bagarius yarrelli (Sykes, 1839)	Freshwater		Capture Fishery	200 cm SL	NT	Chandrashekhariah *et al.* (2000)
Gagata cenia (Hamilton, 1822)	Freshwater; brackish		Capture Fishery	15.0 cm SL	LC	Naik *et al.* (2013c); Naik *et al.* (2013d)
Gagata itchkeea (Sykes, 1839)	Freshwater		Capture Fishery	7.6 cm	VU	Chandrashekhariah *et al.* (2000); Jen (2002)
Glyptothorax annandalei Hora, 1923	Freshwater	ENI	Fishery doesn't exists	11.5 cm	LC	Jen (2002)
Glyptothorax lonah (Sykes, 1839)	Freshwater	ENI	Fishery doesn't exists	15.0 cm	LC	Chandrashekhariah *et al.* (2000); Ramachandra *et al.* (2012)
Glyptothorax madraspatanus (Day, 1873)	Freshwater	ENCR	Fishery doesn't exists	11.5 cm	EN	Jen (2002)
Glyptothorax trewavasae Hora, 1938	Freshwater	ENWG, ENKR	Fishery doesn't exists	14.0 cm	VU	Chandrashekhariah *et al.* (2000)
Order: Symbranchiformes						
Family: Mastacembelidae						

Order, Family, Scientific name	Environment	Endemic, Exotic	Human Use	Maximum size (TL)	IUCN Status	References
Macrognathus aral (Bloch & Schneider, 1801)	Freshwater; brackish		Capture Fishery	63.5 cm	LC	Chandrashekhariah *et al.* (2000)
Macrognathus pancalus Hamilton, 1822	Freshwater; brackish		Capture Fishery	18.0 cm	LC	Chandrashekhariah *et al.* (2000)
Mastacembelus armatus (Lacepède, 1800)	Freshwater; brackish		Ornamental	90.0 cm	LC	Chandrashekhariah *et al.* (2000); Naik *et al.* (2013c); Naik *et al.* (2013d); Naik *et al.* (2014)
Order: Syngnathiformes						
Family: Syngnathidae						
Microphis cuncalus (Hamilton, 1822)	Freshwater; brackish		Fishery doesn't exists	17.5 cm SL	LC	Venkateshwarlu *et al.* (2014)
Order: Tetraodontiformes						
Family: Tetraodontidae						
Carinotetraodon imitator Britz & Kottelat, 1999	Freshwater	ENKL	Fishery doesn't exists	-	DD	Anandhi & Sharath (2014)
Carinotetraodon travancoricus (Hora & Nair, 1941)	Freshwater	ENI	Ornamental	3.5 cm	VU	Venkateshwarlu *et al.* (2014)

ENI = endemic to India; ENPI = endemic to peninsular India; ENSI = endemic to south India; ENWG = endemic to Western Ghats of India; ENCR = endemic to Cauvery River; ENKR = endemic to Krishna River; ENSR = endemic to Sharavati River; ENMH = endemic to Maharashtra; ENKA = endemic to Karnataka; ENKL = endemic to Kerala; ENTN = endemic to Tamil Nadu; EX – Exotic; TL = Total Length; SL = Standard Length; IUCN - International Union for Conservation of Nature; CR – Critically Endangered; EN – Endangered; VU – Vulnerable; NT – Near Threatened; LC – Least Concern; DD – Data Deficient; NE – Not Evaluated

Table 2. Number of family, genera and species under each Order of Freshwater Fishes of Karnataka

Order	Family	Number of Genera	Number of Species
Angulliformes	Anguillidae	1	2
Beloniformes	Adrianichthyidae	1	2
	Belonidae	2	2
	Hemiramphidae	1	2
Clupiformes	Clupeidae	1	1
	Engraulidae	1	1
Cypriniformes	Balitoridae	2	2
	Cobitidae	2	3
	Cyprinidae	35	123
	Nemacheilidae	5	15
	Psilorhynchidae	1	1
Cyprinodontiformes	Aplocheilidae	1	1
	Poeciliidae	1	1
Elopiformes	Megalopidae	1	1
Gonorynchiformes	Chanidae	1	1
Mugiliformes	Mugilidae	1	1
Osteoglossiformes	Notopteridae	2	2
Perciformes	Ambassidae	3	5
	Anabantidae	1	1
	Badidae	1	1
	Channidae	1	5
	Cichlidae	2	5

Order	Family	Number of Genera	Number of Species
	Gerreidae	1	3
	Gobidae	2	2
	Nandidae	1	1
	Osphronemidae	3	3
	Pristolepididae	1	1
Siluriformes	Amblycipitidae	1	1
	Bagridae	6	20
	Claridae	1	3
	Heteropneustidae	1	1
	Loricardiidae	1	1
	Pangasiidae	1	1
	Schilbeidae	7	10
	Siluridae	3	6
	Sisoridae	3	8
Symbranchiformes	Mastacembelidae	2	3
Syngnathiformes	Syngnathidae	1	1
Tetraodontiformes	Tetraodontidae	1	2
Total = 14	**Total** = 39	**Total** = 103	**Total** = 245

Figure 1. Family wise representation of freshwater fishes of Karnataka

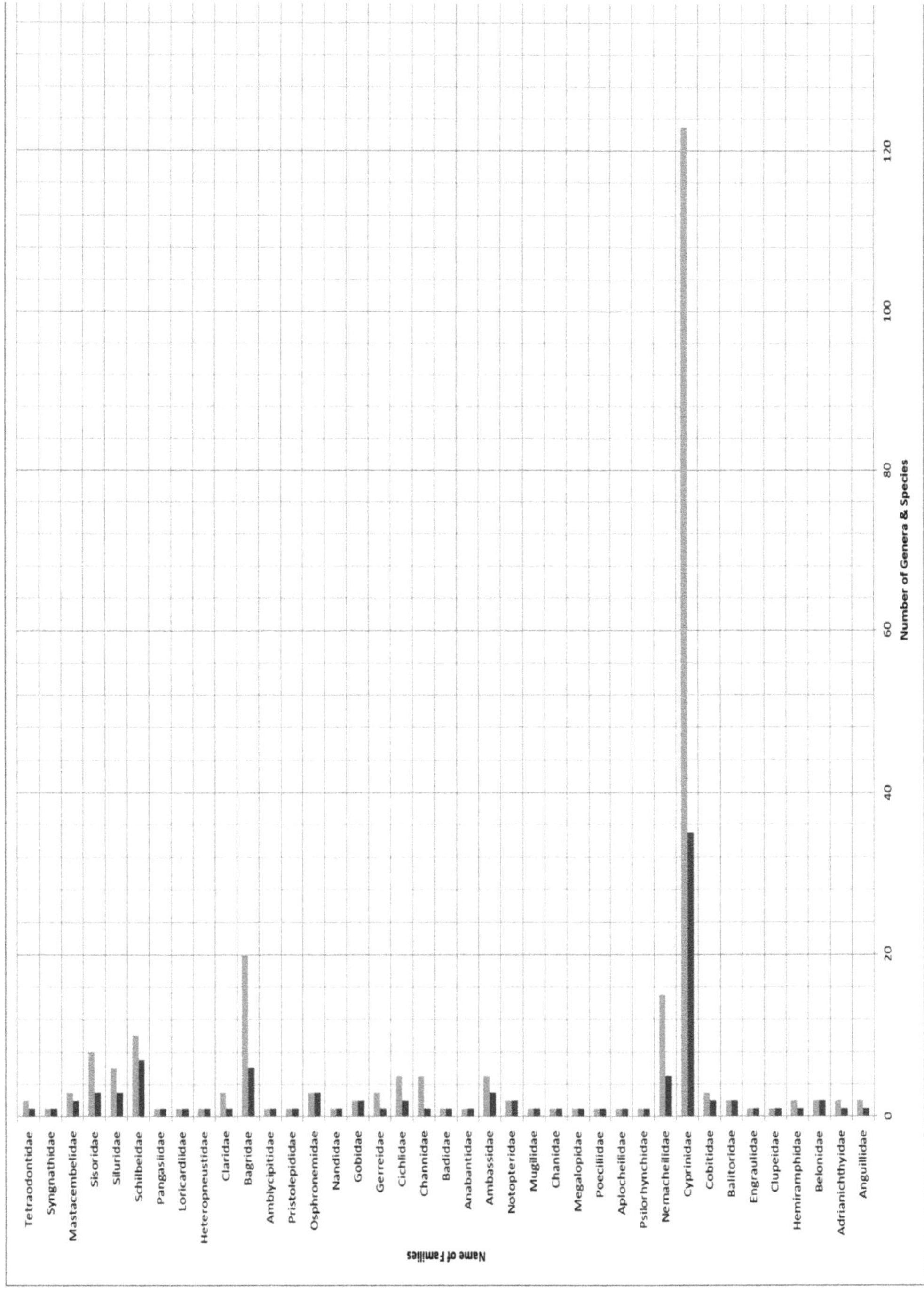

3

FRESHWATER FISH DIVERSITY OF KERALA

H.S. Mogalekar[1], *P. Jawahar*[1],
C.P. Ansar[2] *and Prateek*[1]

[1] *Department of Fisheries Biology and Resource Management, Fisheries College and Research Institute, Tamil Nadu Fisheries University, Thoothukudi-628 008, Tamil Nadu, India*

[2] *Regional Agriculture Research Station, Kerala Kerala Agricultural University, Kumarokom- 686 563, Kerala, India*

INTRODUCTION

Kerala is a blessed with rich freshwater resources which harbour a diversified fish fauna characterized by many endemic fish species. The fast flowing streams and rivers have been an excellent habitat and environment enabling evolution of rich fish diversity. The total area of freshwater resources in Kerala is about 1, 35,568 ha. There are 44 rivers in the state with 3200 km of total length and 85000 ha of water spread area. The 41 rivers are west flowing, most of them having their origin in the Western Ghats and draining into the Arabian Sea. The major west flowing rivers of the state are Bharathapuzha, Periyar, Pampa, Menachil, Chalakudypuzha, Achankovil, Manimala, Neyar, Chandragiri, Mahipuzha and Anjarakandi. The other three rivers, Kabini, Bhavani and the Pambar also originate in the Western Ghats but are east flowing. Other freshwater resources include 46173 ponds with total water spread area of 10584 ha, 4731 freshwater farms with total area of 733 ha, 9 fresh water lakes with total water spread area of 1620 ha, 70 number of bunds/ barriers/anicut/shelter water holds with total area of 879 ha, 80 check dams with total water spread area of 259 ha, 1051 irrigation tankslakes with total water spread area of 2313.27 ha, 49 reservoirs with total water spread area of 34180 ha, 7 freshwater springs and 11 water falls (Kerala Inland Fisheries Statistics, 2016).

The comprehensive studies on the freshwater fish fauna of Kerala are those of Day (1865, 1878, 1889); Pillai (1929); John (1936); Hora and Law (1941); Silas (1951a, 1951b); Remadevi and Indra (1986); Pethiyagoda and Kottelat (1994); Kurup (1994); Kurup and Ranjeet (2002); Easa and Shaji (1995); Menon and Jacob (1996); Manimekalan and Das (1998); Ajithkumar *et al.* (1999); Arun (1999); Biju (1999); Thomas *et al.* (1999a and b); Euphrasia (2004); Kurup *et al.* (2004); Beevi and Ramachandran (2009); Johnson and Arunachalam (2009); Baby *et al.* (2010); Binoy (2010); Radhakrishnan and Kurup (2010); Abraham *et al.* (2011); Renjithkumar *et al.* (2011); Bijukumar *et al.* (2013); Raghavan *et al.* (2013); Ambili *et al.* (2014); Bijukumar and Raghavan (2015); Plamoottil (2015). Meanwhile numerous changes in taxonomy and / or systematic placement of freshwater fish species of Kerala took place. Generation of proper database on fish germplasm resources is vital for management, conservation and to evolve a plan for proper utilization of fishery resources. Hence this restructured information on species composition of freshwater fishes from various localities of Kerala is document in this paper.

MATERIALS AND METHODS

The checklist of the freshwater fishes of Kerala was prepared based on an exhaustive review of published literature, short notes and reports, research articles, taxonomic assessments and previous species checklists from Kerala. Taxonomic treatment is based on Eschmeyer *et al.* (2016) and Froese & Pauly (2013). Details on the environment, endemic or exotic status and human use (existence or non existence of fishery, aquaculture, ornamental or gamefish) of all species were obtained by retrieving species level information from the published literature as well as FishBase (Froese & Pauly, 2013). The endemicity or exoticity of all species indicated based on Indian subcontinent. The listing of food fishes is based on the maximum size of the species, acceptance by local people and discussion with fishermen. The list of ornamental fishes is prepared based on colouration pattern, shape and maximum size (Table 1). Information on the conservation status of all taxa in this paper was retrieved from the International Union for Conservation of Nature (IUCN) Red List of Threatened Species, whose underlying assessments are based on the IUCN Red List Categories and criteria (Version 2015.2) downloaded on 05 October 2015 (IUCN, 2015).

RESULTS

The present checklist revealed occurrence of 273 species of fishes belonging to 16 orders, 45 families and 106 genera from the freshwater bodies of Kerala (Table 2). These included 196 truly freshwater species, 54 freshwater cum brackish species and 23 marine cum freshwater cum brackish species. Freshwater fishes of Kerala consist of 113 species with ornamental value, 44 species forms candidate species for aquaculture, 119 species are important for capture fishery and 26 species are important for game fishing or sport fishing (Table 1). The top three orders with diverse species composition were Cypriniformes (154 species, 44 genera and 4 families), Siluriformes (51 species, 19 genera and 10 families) and Perciformes (37 species, 20 genera and 13 families) (Table 2). The most diverse family was Cyprinidae with a total of 121 species from 32 genera (44.32 %) followed by Bagridae with 23 species and 4 genera and Nemacheilidae with 19 species and 5 genera (Table 2). out of the 273 species, 15 species are exotic and alien, 60 species are endemic to Kerala, 33 species are endemic to Western Ghats of India, 5 species are endemic to Peninsular India, 4 species are endemic to South India, 6 species are endemic to Tamil Nadu, 9 species are endemic to Cauvery River and 46 species are endemic to India. *Nemacheilus triangularis, Mystus oculatus* and *Clarias dayi* are endemic to Kerala and Tamil Nadu. Species endemic to Tamil Nadu reported from Kerala are *Nemacheilus nilgiriensis, Nemacheilus triangularis, Dawkinsia rohani, Dawkinsia tambraparniei, Mystus oculatus* and *Clarias dayi* (Table 1).

Among the species listed under threatened category, 6 were critically endangered while 44 species were endangered, whereas 21 species were vulnerable. Whereas 202 species were under the non-threatened category, among which 9 species were nearly threatened, 137 species were least concern, 14 species were data deficient and 42 species have not been evaluated for IUCN status (Table 1). Species such as *Mesonoemacheilus herrei, Barbodes wynaadensis, Horalabiosa arunachalami, Pethia pookodensis, Puntius thomassi* and *Hemibagrus punctatus* are critically endangered. The endangered category includes *Homaloptera montana, Homaloptera santhamparaiensis, Travancoria elongata, Travancoria jonesi, Botia striata, Longischistura striata, Nemacheilus petrubanarescui, Nemacheilus pulchellus, Barilius canarensis, Crossocheilus periyarensis, Devario neilgherriensis, Garra hughi, Garra surendranathanii, Horalabiosa joshuai, Lepidopygopsis typus, Hypselobarbus dubius, Hypselobarbus micropogon, Hypselobarbus periyarensis, Labeo potail, Hypselobarbus curmuca, Osteochilus longidorsalis, Dawkinsia arulius, Dawkinsia exclamatio, Dawkinsia tambraparniei, Puntius cauveriensis, Puntius crescentus, Puntius ophicephalus, Sahyadria chalakkudiensis, Sahyadria denisonii, Tor khudree, Tor malabaricus, Hypselobarbus mussullah, Tor putitora, Horabagrus nigricollaris, Pangasianodon hypophthalmus, Pseudeutropius mitchelli, Silonia children, Pterocryptis wynaadensis, Glyptothorax anamalaiensis, Glyptothorax davissinghi, Glyptothorax housei, Glyptothorax madraspatanus, Mastacembelus oatesii* and *Monopterus fossorius*.

DISCUSSION

The available information on the freshwater fishes of Kerala is mostly on systematics, distribution and abundance, henceforward present paper provides an updated checklist with the latest taxonomic revisions and range extensions, together with their environment, endemic and exotic status, maximum size, fisheries status and conservation status as per IUCN criteria (Table 1). We made the best use of available information on fish fauna in developing this checklist and hope it will be assist to policy makers, managers and conservationists.

All the following reports agreed with dominance of Cypriniformes, Siluriformes and Perciformes over other orders of freshwater fishes. The Cyprinidae is the most diverse group among the freshwater fishes and many species form commercially important fisheries and most of interest to aquaculture.

Easa and Basha (1995) listed 91 species belonging to 24 families and 46 genera from Nilgiri Biosphere Reserve which includes 58 species belonging to 17 families and 31 genera from Kabani river and Vythiripuzha in Wayanad and 50 species of 34 genera belonging to 21 families from Chaliyar river system. Arun (1999) reported 27 fish species from Periyar Lake in Kerala. Euphrasia (2004) reported 122 fish species from 19 rivers of Kerala. Kurup *et al.* (2004) documented 175 fish species under 13 orders, 29 families and 65 genera from the 41 west flowing and 3 east flowing river systems of Kerala. Raghavan *et al.* (2008) reported 71 species of freshwater fishes belonging to 27 families and 50 genera from Chalakudy river of Kerala. Beevi and Ramachandran (2009) recorded 69 freshwater fishes belonging to 20 families and 7 orders from Ernakulam District of Kerala. Johnson and Arunachalam (2009) recorded 60 species of primary freshwater fishes belonging to four orders, 13 families and 27 genera were from the streams of southern Western Ghats. Baby *et al.* (2010) recorded 43 species of freshwater fish belonging to 13 families and 28 genera from various tributaries of Chaliyar River draining of Kerala. Radhakrishnan and Kurup (2010) reported 54 fish species belonging to six orders and 19 families from Periyar Tiger Reserve of Kerala. Abraham *et al.* (2011) reported 103 fish species belonging to 53 genera and 24 families from west- owing drainages of the Ashambu (Agasthyamala) Hills landscape in the southern Western Ghats of Kerala. Renjithkumar *et al.* (2011) recorded 60 species of fishes belonging to 17 families and 36 genera from the Pampa River of Kerala. Bijukumar *et al.* (2013) recorded 117 species under 42 families and 81 genera from the Bharathapuzha River system of Kerala. Plamoottil (2015) recorde 97 species of fishes including 12 new fish recordes from Manimala River of Kerala.

The available literature pointed out the main cause for the alarming decline of fish populations in most of the freshwater bodies of Kerala are anthropogenic activities such as unsustainable and unethical fishing by using fish poisons, dynamite fishing, sand-mining and introduction of invasive exotic species. Habitat destruction of natural spawning and breeding grounds of the fishes through sand extraction and construction of physical obstructions across rivers has contributed to the population decline and the endangerment of the freshwater fishes. Existing conservation policies need to be implementd to reduce species extinction, as 71 species of freshwater fishes listed under threatened category. As numbers of species for capture fishery, aquaculture, ornamental and game fishing or sport fishing are high, seed production by selective breeding is recommended for income generation through culture of these species. Selective breeding and ranching of native fish species might help to overcome the problems of species endangerment.

CONCLUSIONS

Freshwater bodies of Kerala have exceptional fish biodiversity with a high degree of endemism. Abiding management plans are required to conserve and preserve the fish germplasm of Kerala, as 26.0% of fishes are coming under threatened catagory.

REFERENCES

Abraham, R.K., Kelkar, N. & Bijukumar, A. 2011. Freshwater fish fauna of the Ashambu Hills landscape, southern Western Ghats, India, with notes on some range extensions. Journal of Threatened Taxa, 3(3): 1585-1593.

Ajithkumar, C.R., Remadevi, K., Thomas, R.K. & Biju, C.R. 1999. Fish fauna, abundance and distribution in Chalakudy river system, Kerala. J. Bombay Nat. Hist. Soc. 96(2): 244-254.

Ambili, T.R., Manimekalan, A. & Verma, M.S. 2014. Genetic diversity of genus Tor in river Chaliyar, Southern Western Ghats, Kerala: Through DNA barcoding. Journal of Science, 4(4): 206-214.

Arun, L.K. 1999. Patterns and processes of fish assemblages in Periyar Lake Valley system of Southern Western Ghats. Kerala forest research institute peechi, Thrissur, Research Report 172, Pages: 57.

Baby, F., Tharian, J., Ali, A. & Raghavan, R. 2010. A checklist of freshwater fishes of the New Amarambalam Reserve Forest (NARF), Kerala, India. Journal of Threatened Taxa, 2(12): 1330-1333.

Beevi, K.S.J. & Ramachandran, A. 2009. Checklist of freshwater fishes collected from Ernakulam District, Kerala, India. Journal of Threatened Taxa, 1(9): 493-494.

Biju, C.R., Thomas, R. & Ajithkumar, C.R., 1999. Fishes of Parambikulam Wildlife Sanctuary, Palakkad District, Kerala. J. Bombay. Nat. Hist. Soc. 96(1): 82-87.

Bijukumar, A. & Raghavan, R. 2015. A checklist of fishes of Kerala, India. Journal of Threatened Taxa, 7(13): 8036–8080 http://dx.doi.org/10.11609/jott.2004.7.13.8036-8080.

Bijukumar, A., Philip, S., Ali, A., Sushama, S. & Raghavan, R. 2013. Fishes of River Bharathapuzha, Kerala, India: diversity, distribution, threats andconservation. Journal of Threatened Taxa, 5(15): 4979-4993 http://dx.doi.org/10.11609/jott.o3640.4979-93.

Binoy, V. V. 2010. Catfish Clarias is vanishing from the waters of Kerala. Current Science, 99(6): 713-714.

Day, F. 1865. The fishes of Malabar. London, Bernard Quaritch.

Day, F.1878. The fishes of India, being a natural history of the fishes known to inhabit the seas and freshwaters of India, Burma and Ceylon. Dawson, London 1: 778; 2:195 plates reprint 1958.

Day, F.1889. Fauna of British India, Including Ceylon and Burma. Fishes 1. 548 pp. 2: 509. London, Taylor and Francis.

Easa, P.S. & Basha, C.C. 1995. A survey on the habitat and distribution of stream fishes in the kerala part of nilgiri biosphere reserve. Kerala Forest Research Institute Peechi, Thrissur, KFRI Research Report 104. Pp 1-84.

Easa, P.S. & Shaji, C.P., 1995. Freshwater fish diversity in Kerala part of the Nilgiri Biosphere Reserve. Research Report. Peechi, Kerala Forest Research Institute.

Eschmeyer, W.N., Fricke, R. & van der Laan, R. (Eds). 2016. Catalog of fishes: genera, species, references. Electronic version accessed 19-28 March 2016.(http://researcharchive.calacademy.org/research/ichthyology/catalog/fishcatmain.asp). [This version was edited by Bill Eschmeyer.]

Euphrasia, C.J. 2004. Bionomics, resource characteristics and distribution of the threatened freshwater fishes of Kerala. Ph.D. Thesis submitted to the Cochin University of Science and Technology. Pp 453.

Froese, R. and Pauly, D. (Eds.) 2013. List of Freshwater Fishes reported from India. In: FishBase. www.fishbase.org, 23 October 2015.

Hora, S.L. & Law, N.C. 1941. The freshwater fishes of Travancore. Rec. Ind. Mus. 43: 233-256.

IUCN, 2015. The IUCN Red List of Threatened Species. Version2015.2. IUCN, Gland, Switzerland and Cambridge, UK, http://www.iucnredlist.org. 23 October 2015.

John, C.C. 1936. Freshwater fish and fisheries of Travancore. J. Bombay Nat. Hist. Soc. XXXV: 132-157

Johnson, J.A & Arunachalam, M. 2009. Diversity, distribution and assemblage structure of fishes in streams of southern Western Ghats, India. Journal of Threatened Taxa, 1(10): 507-513.

Kerala Inland Fisheries Statistics, 2016. Government of Kerala, Department of Fisheries. http://www.fisheries.kerala.gov.in/index.php?option=com_content&view=article&id=77:inlandfisheries&catid=41:inlandfisheries&Itemid=45 (3/12/2016)

Kurup, B.M. & Ranjeet, K. 2002. Invasion of exotic fish population in Periyar lake, Kerala: Ahotspot of fish biodiversity.In Proc. Life History Traits of Freshwater Fish Population for its Utilization in Conservation. Lucknow, India, NBFGRNATP, AC-15. Pp.1–4.

Kurup, B.M. 1994. An account on threatened fishes of river systems flowing through Kerala. In Proc. Nat. Sem. Endangered Fish India. pp. 129-140

Kurup, B.M., Radhakrishnan, K.V. & Manojkumar, T.G. 2004. Biodiversity status of fishes inhabiting rivers of kerala (s. India) with special reference to endemism, threats and conservation measures. In: Welcomme, R.L. and Petr, T. (Eds) 2004. Proceedings of the second international symposium on the management of large rivers for fisheries (Volume II), Sustaining Livelihoods and Biodiversity in the New Millennium, 11 – 14 February 2003, Phnom Penh, Kingdom of Cambodia, Pp 163-182.

Manimekalan, A. & Das, H.S. 1998. Glyptothorax davissinghi Pisces: Sisoridae, a new catfish from Nilambur in the Nilgiri Biosphere Reserve, South India. J. Bombay Nat. Hist. Soc. 95: 8791.

Menon, A.G.K. & Jacob, P.C. 1996. Crossocheilus periyarensis, a new cyprinid fish from Thannikkudy Thekkadi, Kerala, S. India. J. Bombay Nat. Hist. Soc. 95: 87-91.

Pethiyagoda, R. & Kottelat, M. 1994. Three new species of fishes of the genera Osteochilichthys Cyprinidae, Travancoria Balitoridae and Horabagrus Bagridae from the Chalakkudy river, Kerala, India. Journal of South Asian Natural History. 11: 97-116.

Pillai, R.S.N. 1929. A list of fishes taken in Travancore from 1901-1915. J. Bombay Nat. Hist. Soc. XXX: 111-126.

Plamoottil, M. 2015. Ichthyodiversity of Manimala River, Kerala, India. Journal of Zoological and Bioscience Research, 2(2):26-34.

Radhakrishnan, K.V. & Kurup, B.M. 2010. Ichthyodiversity of Periyar Tiger Reserve, Kerala, India. Journal of Threatened Taxa, 2(10): 1192-1198.

Raghavan, R., Prasad, G., Ali, A.P.H. & Pereira, B. 2008. Fish fauna of Chalakudy River, part of Western Ghats biodiversity hotspot, Kerala, India: patterns of distribution, threats and conservation needs. Biodivers Conserv (2008), 17:3119–3131.

Raghavan, R., Tharian, J., Ali, A., Jadhav, S. & Dahanukar, N. 2013. Balitora jalpalli, A new species of Stone loach (Teleostei: Cypriniformes: Balitoridae) from Silent Valley, Southern Western Ghats, India. Journal of Threatened Taxa, 5(5): 3921–3934.

Remadevi, K. & Indra, T.J. 1986. Fishes of Silent Valley. Rec. Zool. Surv. India. 841(4): 243-257.

Renjithkumar, C. R., Harikrishnan, M. & Kurup, B. M. 2011. Exploited fisheries resources of the Pampa River, Kerala, India. Indian J. Fish., 58(3): 13-22.

Seethal, L.S., Jaya, D.S. & Williams, S.E. 2014. Ichthyofaunal diversity of Vattakkayal, a part of Ashtamudi lake, Kollam district, Kerala, South India. Journal of Aquatic Biology and Fisheries, 2/2014/ pp. 620 to 626.

Silas E.G. 1951a. On a collection of fishes from the Anamalai and Nelliampathy hill ranges Western Ghatswith: Notes on its zoogeographical significance. J. Bombay Nat. Hist. Soc. 49: 670-681.

Silas E.G. 1951b. Fishes from the high ranges of Travancore. J. Bombay Nat. Hist. Soc. 502: 323-330.

Thomas, R.K., Biju, C.R. & Ajithkumar, C.R. 1999a. Extension of range of Esomus thermoicos Pisces: Cyprinidae: Rasborinaeto Kerala. J. Bombay Nat. Hist. Soc., 96(1): p. 163.

Thomas, R.K.,, Biju, C.R. & Ajithkumar, C.R. 1999b. Distribution of Pangio goaensis TilakCypriniFormes: Cobitidae in Manimala River, Southern Kerala. J. Bombay Nat. Hist. Soc. 96(3): 479.

Table 1. Freshwater Fishes of Kerala with note on environment, endemic or exotic status, maximum size, human use and conservation status

Order, Family, Scientific name	Environment	Endemic, Exotic	Maximum size (TL)	Human Use	IUCN Status	References
Order: Anguilliformes						
Family: Anguillidae						
Anguilla bengalensis (Gray, 1831)	Marine; freshwater; brackish		200 cm	Capture Fishery; Aquaculture; Gamefish	NT	Easa & Basha (1995); Euphrasia (2004); Kurup *et al.* (2004); Radhakrishnan & Kurup (2010); Abraham *et al.* (2011); Bijukumar *et al.* (2013); Bijukumar & Raghavan (2015); Plamoottil (2015)
Anguilla bicolor McClelland, 1844	Marine; freshwater; brackish		123 cm	Capture Fishery	NT	Euphrasia (2004); Kurup *et al.* (2004); Raghavan *et al.* (2008); Abraham *et al.* (2011); Bijukumar *et al.* (2013); Bijukumar & Raghavan (2015); Plamoottil (2015)
Family: Ophichthidae						
Pisodonophis boro (Hamilton, 1822)	Marine; freshwater; brackish		100.0 cm	Capture Fishery	LC	Kurup *et al.* (2004)
Order: Beloniformes						
Family: Belonidae						
Strongylura strongylura (van Hasselt, 1823)	Marine; freshwater; brackish		40.0 cm SL	Capture Fishery	NE	Beevi & Ramachandran (2009); Bijukumar & Raghavan (2015); Plamoottil (2015)

Order, Family, Scientific name	Environment	Endemic, Exotic	Maximum size (TL)	Human Use	IUCN Status	References
Xenentodon cancila (Hamilton, 1822)	Freshwater		40.0 cm	Ornamental	LC	Easa & Basha (1995); Beevi & Ramachandran (2009); Radhakrishnan & Kurup (2010); Abraham *et al.* (2011); Renjithkumar *et al.* (2011); Bijukumar *et al.* (2013); Bijukumar & Raghavan (2015); Plamoottil (2015)
Family: Hemiramphidae						
Hyporhamphus limbatus (Valenciennes, 1847)	Marine; freshwater; brackish		35.0 cm	Capture Fishery	LC	Easa & Basha (1995); Beevi & Ramachandran (2009); Abraham *et al.* (2011); Bijukumar *et al.* (2013); Plamoottil (2015)
Hyporhamphus xanthopterus (Valenciennes, 1847)	Marine; freshwater; brackish		15.0 cm	Ornamental	VU	Beevi & Ramachandran (2009); Bijukumar *et al.* (2013); Plamoottil (2015)
Order: Clupeiformes						
Family: Clupeidae						
Dayella malabarica (Day, 1873)	Marine; freshwater; brackish		6.0 cm SL	Capture Fishery	LC	Euphrasia (2004); Kurup *et al.* (2004); Raghavan *et al.* (2008); Beevi & Ramachandran (2009); Abraham *et al.* (2011); Bijukumar *et al.* (2013); Plamoottil (2015)
Order: Cypriniformes						
Family: Balitoridae						
Balitora brucei Gray, 1830	Freshwater		10.5 cm SL	Fisheries: of no interest	NT	Euphrasia (2004); Kurup *et al.* (2004)
Balitora jalpalli Raghavan, Tharian, Ali, Jadhav & Dahanukar, 2013	Freshwater	ENKL	6.2 cm SL	Fisheries: of no interest	NE	Bijukumar *et al.* (2013); Raghavan *et al.* (2013); Bijukumar & Raghavan (2015)

Order, Family, Scientific name	Environment	Endemic, Exotic	Maximum size (TL)	Human Use	IUCN Status	References
Balitora mysorensis Hora, 1941	Freshwater	ENWG	5.0 cm SL	Fisheries: of no interest	VU	Easa & Basha (1995); Euphrasia (2004); Kurup *et al.* (2004); Bijukumar & Raghavan (2015);
Bhavania australis (Jerdon, 1849)	Freshwater	ENWG	9.0 cm SL	Ornamental	LC	Easa & Basha (1995); Arun (1999); Euphrasia (2004); Kurup *et al.* (2004); Raghavan *et al.* (2008); Johnson & Arunachalam (2009); Baby *et al.* (2010); Radhakrishnan & Kurup (2010); Abraham *et al.* (2011); Bijukumar *et al.* (2013); Bijukumar & Raghavan (2015); Plamoottil (2015)
Homaloptera menoni Shaji & Easa, 1995	Freshwater	ENI	-	Fisheries: of no interest	LC	Easa & Basha (1995); Kurup *et al.* (2004); Bijukumar *et al.* (2013); Bijukumar & Raghavan (2015)
Homaloptera pillaii Indra & Devi, 1981	Freshwater	ENKL	7.5 cm SL	Fisheries: of no interest	LC	Easa & Basha (1995); Kurup *et al.* (2004); Bijukumar *et al.* (2013); Bijukumar & Raghavan (2015)
Homaloptera montana Herre, 1945	Freshwater	ENWG	7.3 cm SL	Fisheries: of no interest	EN	Easa & Basha (1995); Kurup *et al.* (2004); Raghavan *et al.* (2008); Bijukumar & Raghavan (2015)
Homaloptera santhamparaiensis Arunachalam, Johnson & Devi, 2002	Freshwater	ENWG	6.1 cm SL	Fisheries: of no interest	EN	Johnson & Arunachalam (2009); Bijukumar & Raghavan (2015)
Homaloptera silasi Kurup & Radhakrishnan, 2011	Freshwater	ENKL	-	Fisheries: of no interest	NE	Bijukumar & Raghavan (2015)
Travancoria elongata Pethiyagoda & Kottelat, 1994	Freshwater	ENKL	11.4 cm	Fisheries: of no interest	EN	Euphrasia (2004); Kurup *et al.* (2004); Raghavan *et al.* (2008); Bijukumar & Raghavan (2015)
Travancoria jonesi Hora, 1941	Freshwater	ENKL	8.4 cm SL	Ornamental	EN	Arun (1999); Euphrasia (2004); Kurup *et al.* (2004); Radhakrishnan & Kurup (2010); Abraham *et al.* (2011); Bijukumar & Raghavan (2015)

Order, Family, Scientific name	Environment	Endemic, Exotic	Maximum size (TL)	Human Use	IUCN Status	References
Family: Cobitidae						
Botia striata Narayan Rao, 1902	Freshwater	ENWG	7.8 cm SL	Ornamental	EN	Johnson & Arunachalam (2009)
Lepidocephalichthys thermalis (Valenciennes, 1846)	Freshwater		38.0 cm SL	Ornamental	LC	Easa & Basha (1995); Arun (1999); Kurup *et al.* (2004); Beevi & Ramachandran (2009); Raghavan *et al.* (2008); Baby *et al.* (2010); Radhakrishnan & Kurup (2010); Abraham *et al.* (2011); Bijukumar *et al.* (2013); Bijukumar & Raghavan (2015)
Pangio goaensis (Tilak, 1972)	Freshwater	ENWG	3.1 cm SL	Ornamental	LC	Kurup *et al.* (2004); Abraham *et al.* (2011); Bijukumar *et al.* (2013); Bijukumar & Raghavan (2015)
Family: Nemacheilidae						
Acanthocobitis botia (Hamilton, 1822)	Freshwater		11.0 cm	Fisheries: of no interest	LC	Euphrasia (2004); Kurup *et al.* (2004)
Acanthocobitis mooreh (Sykes, 1839)	Freshwater	ENPI	4.4 cm SL	Fisheries: of no interest	LC	Easa & Basha (1995); Bijukumar & Raghavan (2015)
Indoreonectes evezardi (Day, 1872)	Freshwater	ENWG	3.8 cm SL	Fisheries: of no interest	LC	Kurup *et al.* (2004)
Longischistura striata (Day, 1867)	Freshwater	ENWG	5.0 cm SL	Fisheries: of no interest	EN	Kurup *et al.* (2004); Bijukumar & Raghavan (2015)
Mesonoemacheilus herrei Nalbant & B n rescu, 1982	Freshwater	ENKL	-	Fisheries: of no interest	CR	Bijukumar & Raghavan (2015)
Mesonoemacheilus pambarensis (Rema Devi & Indra, 1994)	Freshwater	ENKL	-	Fisheries: of no interest	VU	Euphrasia (2004); Kurup *et al.* (2004); Bijukumar & Raghavan (2015)

Order, Family, Scientific name	Environment	Endemic, Exotic	Maximum size (TL)	Human Use	IUCN Status	References
Mesonoemacheilus remadevii Shaji, 2002	Freshwater	ENKL	5.9 cm SL	Fisheries: of no interest	LC	Bijukumar *et al.* (2013); Bijukumar & Raghavan (2015)
Nemacheilus anguilla Annandale, 1919	Freshwater	ENWG	5.0 cm SL	Fisheries: of no interest	LC	Bijukumar & Raghavan (2015)
Nemacheilus denisoni Day, 1867	Freshwater	ENI	5.0 cm SL	Fisheries: of no interest	LC	Easa & Basha (1995); Arun (1999); Euphrasia (2004); Kurup *et al.* (2004); Johnson & Arunachalam (2009); Radhakrishnan & Kurup (2010); Abraham *et al.* (2011); Bijukumar *et al.* (2013); Bijukumar & Raghavan (2015)
Nemacheilus guentheri Day, 1867	Freshwater	ENPI	5.6 cm SL	Ornamental	LC	Easa & Basha (1995); Arun (1999); Euphrasia (2004); Raghavan *et al.* (2008); Johnson & Arunachalam (2009); Baby *et al.* (2010); Radhakrishnan & Kurup (2010); Bijukumar *et al.* (2013); Bijukumar & Raghavan (2015)
Nemacheilus keralensis (Rita & Nalbant, 1978)	Freshwater	ENKL	5.3 cm SL	Ornamental	VU	Arun (1999); Euphrasia (2004); Kurup *et al.* (2004); Johnson & Arunachalam (2009); Radhakrishnan & Kurup (2010); Bijukumar & Raghavan (2015)
Nemacheilus menoni Zacharias & Minimol, 1999	Freshwater	ENI	-	Fisheries: of no interest	VU	Euphrasia (2004); Kurup *et al.* (2004); Radhakrishnan & Kurup (2010); Bijukumar & Raghavan (2015)
Nemacheilus monilis Hora, 1921	Freshwater	ENWG	4.8 cm SL	Fisheries: of no interest	LC	Easa & Basha (1995); Euphrasia (2004); Kurup *et al.* (2004); Bijukumar *et al.* (2013); Bijukumar & Raghavan (2015)
Nemacheilus nilgiriensis Menon 1987	Freshwater	ENTN	5.1 cm SL	Fisheries: of no interest	LC	Easa & Basha (1995); Kurup *et al.* (2004); Bijukumar & Raghavan (2015)
Nemacheilus periyarensis Kurup & Radhakrishnan, 2005	Freshwater	ENKL	-	Fisheries: of no interest	VU	Kurup *et al.* (2004); Radhakrishnan & Kurup (2010); Bijukumar & Raghavan (2015)

Order, Family, Scientific name	Environment	Endemic, Exotic	Maximum size (TL)	Human Use	IUCN Status	References
Nemacheilus petrubanarescui Menon, 1984	Freshwater	ENI	3.5 cm SL	Fisheries: of no interest	EN	Easa & Basha (1995); Kurup *et al.* (2004)
Nemacheilus pulchellus Day, 1873	Freshwater	ENWG	4.6 cm SL	Fisheries: of no interest	EN	Kurup *et al.* (2004); Abraham *et al.* (2011); Bijukumar & Raghavan (2015)
Nemacheilus semiarmatus Day, 1867	Freshwater	ENPI	5.6 cm SL	Capture Fishery; Aquaculture; Ornamental	LC	Easa & Basha (1995); Euphrasia (2004); Kurup *et al.* (2004); Johnson & Arunachalam (2009); Baby *et al.* (2010); Bijukumar *et al.* (2013); Bijukumar & Raghavan (2015)
Nemacheilus triangularis Day, 1865	Freshwater	ENKL, ENTN	5.8 cm SL	Ornamental	LC	Easa & Basha (1995); Euphrasia (2004); Kurup *et al.* (2004); Raghavan *et al.* (2008); Beevi & Ramachandran (2009); Johnson & Arunachalam (2009); Baby *et al.* (2010); Radhakrishnan & Kurup (2010); Abraham *et al.* (2011); Bijukumar *et al.* (2013); Bijukumar & Raghavan (2015); Plamoottil (2015)
Family: Cyprinidae						
Amblypharyngodon melettinus (Valenciennes, 1844)	Freshwater		8.0 cm	Ornamental	LC	Easa & Basha (1995); Euphrasia (2004); Kurup *et al.* (2004); Raghavan *et al.* (2008); Beevi & Ramachandran (2009); Abraham *et al.* (2011); Bijukumar *et al.* (2013); Bijukumar & Raghavan (2015); Plamoottil (2015)
Amblypharyngodon microlepis (Bleeker, 1853)	Freshwater	ENI	10.0 cm	Ornamental	LC	Euphrasia (2004); Kurup *et al.* (2004); Abraham *et al.* (2011); Renjithkumar *et al.* (2011); Bijukumar *et al.* (2013); Bijukumar & Raghavan (2015); Plamoottil (2015)
Amblypharyngodon mola (Hamilton, 1822)	Freshwater		20.0 cm	Fisheries: of no interest	LC	Kurup *et al.* (2004)

Order, Family, Scientific name	Environment	Endemic, Exotic	Maximum size (TL)	Human Use	IUCN Status	References
Barbodes carnaticus (Jerdon, 1849)	Freshwater	ENI	60.0 cm	Capture Fishery; Aquaculture	LC	Easa & Basha (1995); Euphrasia (2004); Kurup *et al.* (2004); Raghavan *et al.* (2008); Johnson & Arunachalam (2009); Baby *et al.* (2010); Abraham *et al.* (2011); Bijukumar *et al.* (2013); Bijukumar & Raghavan (2015)
Barbodes wynaadensis (Day, 1873)	Freshwater	ENI	25.0 cm	Capture Fishery	CR	Easa & Basha (1995); Euphrasia (2004); Kurup *et al.* (2004); Bijukumar & Raghavan (2015)
Barilius bakeri Day 1865	Freshwater	ENKL	15.0 cm	Ornamental	LC	Easa & Basha (1995); Arun (1999); Euphrasia (2004); Kurup *et al.* (2004); Raghavan *et al.* (2008); Beevi & Ramachandran (2009); Johnson & Arunachalam (2009); Baby *et al.* (2010); Radhakrishnan & Kurup (2010); Abraham *et al.* (2011); Bijukumar *et al.* (2013); Bijukumar & Raghavan (2015); Plamoottil (2015)
Barilius barna (Hamilton, 1822)	Freshwater		15.0 cm	Capture Fishery	LC	Kurup *et al.* (2004)
Barilius bendelisis (Hamilton, 1807)	Freshwater		22.7 cm	Capture Fishery	LC	Euphrasia (2004); Kurup *et al.* (2004); Bijukumar *et al.* (2013); Bijukumar & Raghavan (2015)
Barilius canarensis (Jerdon, 1849)	Freshwater	ENWG	15.0 cm	Ornamental	EN	Euphrasia (2004); Kurup *et al.* (2004); Beevi & Ramachandran (2009); Bijukumar & Raghavan (2015)
Barilius gatensis (Valenciennes, 1844)	Freshwater	ENWG	15.0 cm	Ornamental	LC	Easa & Basha (1995); Euphrasia (2004); Kurup *et al.* (2004); Raghavan *et al.* (2008); Beevi & Ramachandran (2009); Johnson & Arunachalam (2009); Baby *et al.* (2010); Radhakrishnan & Kurup (2010); Abraham *et al.* (2011); Bijukumar *et al.* (2013); Bijukumar & Raghavan (2015)

Order, Family, Scientific name	Environment	Endemic, Exotic	Maximum size (TL)	Human Use	IUCN Status	References
Catla catla (Hamilton,1822)	Freshwater; brackish		182 cm	Capture Fishery; Aquaculture; Gamefish	LC	Euphrasia (2004); Kurup *et al.* (2004); Abraham *et al.* (2011); Renjithkumar *et al.* (2011); Bijukumar *et al.* (2013); Plamoottil (2015)
Cirrhinus cirrhosus (Bloch, 1795)	Freshwater; brackish	ENI	100.0 cm SL	Capture Fishery; Aquaculture; Gamefish	VU	Plamoottil (2015)
Cirrhinus mrigala (Hamilton, 1822)	Freshwater		99.0 cm	Capture Fishery; Aquaculture	LC	Euphrasia (2004); Kurup *et al.* (2004); Abraham *et al.* (2011); Bijukumar *et al.* (2013); Plamoottil (2015)
Cirrhinus reba (Hamilton, 1822)	Freshwater		30.0 cm	Capture Fishery	LC	Euphrasia (2004); Kurup *et al.* (2004)
Crossocheilus latius (Hamilton, 1822)	Freshwater; brackish		15.2 cm	Fisheries: of no interest	LC	Kurup *et al.* (2004)
Crossocheilus periyarensis Menon & Jacob, 1996	Freshwater	ENKL	11.5 cm	Fisheries: of no interest	EN	Arun (1999); Euphrasia (2004); Kurup *et al.* (2004); Radhakrishnan & Kurup (2010); Bijukumar & Raghavan (2015)
Ctenopharyngodon idella (Valenciennes, 1844)	Freshwater	EX	150 cm	Capture Fishery; Aquaculture; Gamefish	NE	Kurup *et al.* (2004); Abraham *et al.* (2011)
Cyprinus carpio Linnaeus, 1758	Freshwater; brackish	EX	120 cm	Capture Fishery; Aquaculture; Gamefish; Ornamental	VU	Easa & Basha (1995); Arun (1999); Euphrasia (2004); Kurup *et al.* (2004); Radhakrishnan & Kurup (2010); Abraham *et al.* (2011); Bijukumar *et al.* (2013); Plamoottil (2015)
Danio rerio (Hamilton, 1822)	Freshwater		3.8 cm SL	Ornamental	LC	Easa & Basha (1995); Johnson & Arunachalam (2009); Bijukumar & Raghavan (2015)

Order, Family, Scientific name	Environment	Endemic, Exotic	Maximum size (TL)	Human Use	IUCN Status	References
Devario aequipinnatus (McClelland, 1839)	Freshwater		15.0 cm	Ornamental	LC	Easa & Basha (1995); Arun (1999); Euphrasia (2004); Kurup *et al.* (2004); Beevi & Ramachandran (2009); Johnson & Arunachalam (2009); Radhakrishnan & Kurup (2010); Abraham *et al.* (2011); Bijukumar *et al.* (2013); Bijukumar & Raghavan (2015); Plamoottil (2015)
Devario fraseri (Hora, 1935)	Freshwater	ENI	10.0 cm	Ornamental	VU	Beevi & Ramachandran (2009)
Devario malabaricus (Jerdon, 1849)	Freshwater		12.0 cm	Ornamental	LC	Easa & Basha (1995); Euphrasia (2004); Kurup *et al.* (2004); Raghavan *et al.* (2008); Beevi & Ramachandran (2009); Baby *et al.* (2010); Radhakrishnan & Kurup (2010); Abraham *et al.* (2011); Bijukumar *et al.* (2013); Bijukumar & Raghavan (2015); Plamoottil (2015)
Devario neilgherriensis (Day, 1867)	Freshwater	ENI	10.0 cm	Fisheries: of no interest	EN	Bijukumar & Raghavan (2015)
Esomus barbatus (Jerdon, 1849)	Freshwater	ENSI	12.0 cm	Ornamental	LC	Bijukumar & Raghavan (2015)
Esomus danricus (Hamilton, 1822)	Freshwater; brackish		13.0 cm	Capture Fishery; Ornamental	LC	Easa & Basha (1995); Kurup *et al.* (2004); Raghavan *et al.* (2008); Abraham *et al.* (2011); Bijukumar *et al.* (2013); Bijukumar & Raghavan (2015)
Esomus malabaricus Day, 1867	Freshwater	ENKL	12.5 cm	Fisheries: of no interest	NE	Beevi & Ramachandran (2009)
Esomus thermoicos (Valenciennes, 1842)	Freshwater		12.7 cm	Capture Fishery; Ornamental	LC	Euphrasia (2004); Kurup *et al.* (2004); Abraham *et al.* (2011); Bijukumar & Raghavan (2015)
Garra emarginata Kurup & Radhakrishnan, 2011	Freshwater	ENKL	8.6 cm SL	Fisheries: of no interest	NE	Bijukumar & Raghavan (2015)

Order, Family, Scientific name	Environment	Endemic, Exotic	Maximum size (TL)	Human Use	IUCN Status	References
Garra gotyla gotyla (Gray, 1830)	Freshwater		18.0 cm	Capture Fishery	LC	Kurup *et al.* (2004)
Garra gotyla stenorhynchus Jerdon, 1849	Freshwater	ENWG	15.5 cm SL	Fisheries: of no interest	LC	Easa & Basha (1995); Euphrasia (2004); Beevi & Ramachandran (2009); Johnson & Arunachalam (2009); Baby *et al.* (2010); Bijukumar & Raghavan (2015)
Garra hughi Silas, 1955	Freshwater	ENKL	15.5 cm SL	Ornamental	EN	Euphrasia (2004); Kurup *et al.* (2004); Johnson & Arunachalam (2009); Abraham *et al.* (2011); Bijukumar & Raghavan (2015)
Garra mlapparaensis Kurup & Radhakrishnan, 2011	Freshwater	ENKL	7.5 cm SL	Fisheries: of no interest	NE	Bijukumar & Raghavan (2015)
Garra mcclellandi (Jerdon, 1849)	Freshwater	ENCR	17.4 cm SL	Fisheries: of no interest	LC	Easa & Basha (1995); Arun (1999); Euphrasia (2004); Kurup *et al.* (2004); Johnson & Arunachalam (2009); Abraham *et al.* (2011); Bijukumar & Raghavan (2015)
Garra mullya (Sykes)	Freshwater	ENI	17.0 cm	Fisheries: of no interest	LC	Easa & Basha (1995); Arun (1999); Euphrasia (2004); Kurup *et al.* (2004); Raghavan *et al.* (2008); Beevi & Ramachandran (2009); Johnson & Arunachalam (2009); Baby *et al.* (2010); Radhakrishnan & Kurup (2010); Abraham *et al.* (2011); Bijukumar *et al.* (2013); Bijukumar & Raghavan (2015); Plamoottil (2015)
Garra menoni Rema Devi & Indra, 1984	Freshwater	ENI	6.9 cm SL	Fisheries: of no interest	VU	Easa & Basha (1995); Euphrasia (2004); Kurup *et al.* (2004); Bijukumar *et al.* (2013); Bijukumar & Raghavan (2015)
Garra periyarensis Gopi, 2001	Freshwater	ENKL	16.0 cm SL	Fisheries: of no interest	VU	Euphrasia (2004); Kurup *et al.* (2004); Radhakrishnan & Kurup (2010); Bijukumar & Raghavan (2015)

Order, Family, Scientific name	Environment	Endemic, Exotic	Maximum size (TL)	Human Use	IUCN Status	References
Garra surendranathanii Shaji, Arun & Easa, 1996	Freshwater	ENKL	14.7 cm SL	Fisheries: of no interest	EN	Euphrasia (2004); Kurup *et al.* (2004); Raghavan *et al.* (2008); Radhakrishnan & Kurup (2010); Abraham *et al.* (2011); Bijukumar & Raghavan (2015)
Haludaria fasciata (Jerdon, 1849)	Freshwater	ENSI	-	Ornamental	LC	Euphrasia (2004); Kurup *et al.* (2004); Raghavan *et al.* (2008); Beevi & Ramachandran (2009); Johnson & Arunachalam (2009); Baby *et al.* (2010); Radhakrishnan & Kurup (2010); Abraham *et al.* (2011); Bijukumar *et al.* (2013); Bijukumar & Raghavan (2015); Plamoottil (2015)
Haludaria kannikattiensis (Arunachalam & Johnson, 2003)	Freshwater	ENI	-	Fisheries: of no interest	LC	Johnson & Arunachalam (2009)
Haludaria melanampyx (Day, 1865)	Freshwater	ENWG	7.5 cm	Fisheries: of no interest	NE	Easa & Basha (1995); Arun (1999); Johnson & Arunachalam (2009); Bijukumar & Raghavan (2015)
Horadandia atukorali Deraniyagala, 1943	Freshwater; brackish		3.0 cm	Fisheries: of no interest	LC	Kurup *et al.* (2004); Beevi & Ramachandran (2009); Bijukumar & Raghavan (2015)
Horalabiosa arunachalami Johnson & Soranam, 2001	Freshwater	ENI	-	Fisheries: of no interest	CR	Johnson & Arunachalam (2009); Bijukumar & Raghavan (2015)
Horalabiosa joshuai Silas, 1954	Freshwater	ENI	9.0 cm SL	Fisheries: of no interest	EN	Kurup *et al.* (2004); Johnson & Arunachalam (2009); Abraham *et al.* (2011); Bijukumar *et al.* (2013)
Hypselobarbus curmuca (Hamilton, 1807)	Freshwater	ENI	120 cm	Capture Fishery; Aquaculture	EN	Easa & Basha (1995); Arun (1999); Euphrasia (2004); Kurup *et al.* (2004); Raghavan *et al.* (2008); Beevi & Ramachandran (2009); Johnson & Arunachalam (2009); Baby *et al.* (2010); Radhakrishnan & Kurup (2010); Abraham *et al.* (2011); Renjithkumar *et al.* (2011); Plamoottil (2015)

Order, Family, Scientific name	Environment	Endemic, Exotic	Maximum size (TL)	Human Use	IUCN Status	References
Hypselobarbus dubius (Day, 1867)	Freshwater	ENI	25.0 cm	Capture Fishery	EN	Euphrasia (2004); Kurup *et al.* (2004); Johnson & Arunachalam (2009)
Hypselobarbus dobsoni (Day, 1876)	Freshwater	ENWG	120 cm	Aquaculture	DD	Johnson & Arunachalam (2009); Bijukumar & Raghavan (2015)
Hypselobarbus jerdoni (Day, 1870)	Freshwater	ENWG	46.0 cm	Capture Fishery	LC	Euphrasia (2004); Kurup *et al.* (2004); Abraham *et al.* (2011); Bijukumar & Raghavan (2015)
Hypselobarbus kolus (Sykes, 1839)	Freshwater	ENI	30.0 cm	Fisheries: of no interest	VU	Euphrasia (2004); Kurup *et al.* (2004); Johnson & Arunachalam (2009); Raghavan *et al.* (2008); Abraham *et al.* (2011)
Hypselobarbus kurali Menon & Rema Devi, 1995	Freshwater	ENI	35.0 cm	Fisheries: of no interest	LC	Kurup *et al.* (2004); Johnson & Arunachalam (2009); Radhakrishnan & Kurup (2010); Abraham *et al.* (2011); Bijukumar *et al.* (2013); Plamoottil (2015)
Hypselobarbus lithopidos (Day, 1874)	Freshwater	ENI	60.0 cm	Capture Fishery	DD	Kurup *et al.* (2004); Bijukumar & Raghavan (2015)
Hypselobarbus micropogon (Valenciennes, 1842)	Freshwater	ENCR	90.0 cm	Capture Fishery; Gamefish	EN	Easa & Basha (1995); Kurup *et al.* (2004); Johnson & Arunachalam (2009); Bijukumar & Raghavan (2015)
Hypselobarbus periyarensis (Raj, 1941)	Freshwater	ENKL	50.0 cm	Capture Fishery	EN	Arun (1999); Euphrasia (2004); Kurup *et al.* (2004); Radhakrishnan & Kurup (2010); Bijukumar & Raghavan (2015)
Bangana ariza (Hamilton, 1807)	Freshwater		30.0 cm SL	Capture Fishery	LC	Easa & Basha (1995); Kurup *et al.* (2004); Plamoottil (2015); Bijukumar & Raghavan (2015)
Labeo calbasu (Hamilton, 1822)	Freshwater; brackish		90.0 cm	Capture Fishery; Aquaculture	LC	Kurup *et al.* (2004); Abraham *et al.* (2011); Bijukumar & Raghavan (2015)

Order, Family, Scientific name	Environment	Endemic, Exotic	Maximum size (TL)	Human Use	IUCN Status	References
Labeo dussumieri (Valenciennes, 1842)	Freshwater		50.0 cm	Capture Fishery; Aquaculture	LC	Euphrasia (2004); Kurup *et al.* (2004); Abraham *et al.* (2011); Renjithkumar *et al.* (2011); Bijukumar & Raghavan (2015); Plamoottil (2015)
Labeo fimbriatus (Bloch, 1795)	Freshwater		91.0 cm	Capture Fishery; Aquaculture	LC	Euphrasia (2004); Kurup *et al.* (2004); Bijukumar *et al.* (2013); Bijukumar & Raghavan (2015)
Labeo kontius (Jerdon, 1849)	Freshwater	ENCR	61.0 cm	Capture Fishery; Aquaculture	LC	Euphrasia (2004)
Labeo potail (Sykes, 1839)	Freshwater	ENCR	30.0 cm	Capture Fishery	EN	Easa & Basha (1995)
Labeo rohita (Hamilton, 1822)	Freshwater		200 cm	Capture Fishery; Aquaculture; Gamefish	LC	Easa & Basha (1995); Euphrasia (2004); Kurup *et al.* (2004); Abraham *et al.* (2011); Bijukumar *et al.* (2013); Plamoottil (2015)
Laubuka dadiburjori Menon, 1952	Freshwater	ENWG	2.5 cm SL	Ornamental	LC	Kurup *et al.* (2004); Beevi & Ramachandran (2009); Abraham *et al.* (2011); Bijukumar *et al.* (2013); Bijukumar & Raghavan (2015)
Laubuka fasciata (Silas, 1958)	Freshwater	ENI	6.0 cm	Fisheries: of no interest	VU	Euphrasia (2004); Kurup *et al.* (2004); Beevi & Ramachandran (2009); Baby *et al.* (2010); Bijukumar *et al.* (2013); Bijukumar & Raghavan (2015)
Laubuka laubuca (Hamilton, 1822)	Freshwater; brackish		7.0 cm	Ornamental	LC	Easa & Basha (1995); Kurup *et al.* (2004); Bijukumar & Raghavan (2015)
Lepidopygopsis typus Raj, 1941	Freshwater	ENKL	25.0 cm	Fisheries: of no interest	EN	Arun (1999); Euphrasia (2004); Kurup *et al.* (2004); Radhakrishnan & Kurup (2010); Bijukumar & Raghavan (2015)
Oreichthys incognito Keith & Kumar 2015	Freshwater	ENKL	-	Fisheries: of no interest	NE	Bijukumar & Raghavan (2015)

Order, Family, Scientific name	Environment	Endemic, Exotic	Maximum size (TL)	Human Use	IUCN Status	References
Osteobrama bakeri (Day, 1873)	Freshwater	ENI	11.0 cm	Ornamental	LC	Easa & Basha (1995); Euphrasia (2004); Kurup *et al.* (2004); Raghavan *et al.* (2008); Abraham *et al.* (2011); Renjithkumar *et al.* (2011); Bijukumar & Raghavan (2015); Plamoottil (2015)
Osteobrama neilli (Day, 1873)	Freshwater	ENI	12.0 cm	Fisheries: of no interest	LC	Bijukumar & Raghavan (2015)
Osteobrama peninsularis Silas, 1952	Freshwater	ENPI	15.0 cm	Capture Fishery	DD	Kurup *et al.* (2004)
Osteochilichthys brevidorsalis (Day, 1873)	Freshwater	ENI	15.0 cm	Fisheries: of no interest	LC	Easa & Basha (1995); Euphrasia (2004); Kurup *et al.* (2004); Bijukumar & Raghavan (2015)
Osteochilichthys nashii (Day, 1869)	Freshwater	ENI	18.0 cm	Capture Fishery	LC	Easa & Basha (1995); Euphrasia (2004); Kurup *et al.* (2004); Johnson & Arunachalam (2009); Baby *et al.* (2010); Bijukumar *et al.* (2013); Bijukumar & Raghavan (2015)
Osteochilichthys thomassi (Day, 1877)	Freshwater	ENWG	32.0 cm	Capture Fishery; Aquaculture	LC	Euphrasia (2004); Kurup *et al.* (2004); Bijukumar & Raghavan (2015)
Osteochilus longidorsalis (Pethiyagoda & Kottelat, 1994)	Freshwater	ENKL	-	Fisheries: of no interest	EN	Euphrasia (2004); Kurup *et al.* (2004); Raghavan *et al.* (2008); Bijukumar *et al.* (2013); Bijukumar & Raghavan (2015)
Dawkinsia arulius (Jerdon, 1849)	Freshwater	ENCR	12.0 cm	Ornamental	EN	Easa & Basha (1995); Euphrasia (2004); Abraham *et al.* (2011)
Dawkinsia assimilis (Jerdon, 1849)	Freshwater	ENI	11.1 cm SL	Ornamental	VU	Bijukumar *et al.* (2013); Bijukumar & Raghavan (2015)
Dawkinsia exclamatio (Pethiyagoda & Kottelat, 2005)	Freshwater	ENKL	8.3 cm SL	Ornamental	EN	Abraham *et al.* (2011); Bijukumar & Raghavan (2015)

Order, Family, Scientific name	Environment	Endemic, Exotic	Maximum size (TL)	Human Use	IUCN Status	References
Dawkinsia filamentosa (Valenciennes, 1844)	Freshwater	ENI	18.0 cm	Capture Fishery; Ornamental	LC	Easa & Basha (1995); Euphrasia (2004); Kurup *et al.* (2004); Raghavan *et al.* (2008); Beevi & Ramachandran (2009); Johnson & Arunachalam (2009); Baby *et al.* (2010); Radhakrishnan & Kurup (2010); Abraham *et al.* (2011); Renjithkumar *et al.* (2011); Bijukumar *et al.* (2013); Bijukumar & Raghavan (2015); Plamoottil (2015)
Dawkinsia rohani (Rema Devi, Indra & Knight, 2010)	Freshwater	ENTN	9.5 cm SL	Ornamental	VU	Bijukumar & Raghavan (2015)
Dawkinsia rubrotinctus (Jerdon, 1849)	Freshwater	ENCR	8.6 cm SL	Fisheries: of no interest	NE	Bijukumar & Raghavan (2015)
Dawkinsia singhala (Duncker, 1912)	Freshwater		9.1 cm SL	Fisheries: of no interest	NE	Kurup *et al.* (2004)
Dawkinsia tambraparniei (Silas, 1954)	Freshwater	ENTN	12.8 cm SL	Ornamental	EN	Johnson & Arunachalam (2009); Abraham *et al.* (2011)
Pethia muvattupuzhaensis (Jameela Beevi & Ramachandran, 2005)	Freshwater	ENKL	4.5 cm SL	Fisheries: of no interest	DD	Beevi & Ramachandran (2009)
Pethia narayani (Hora, 1937)	Freshwater	ENI	7.8 cm	Ornamental	LC	Johnson & Arunachalam (2009); Abraham *et al.* (2011); Bijukumar & Raghavan (2015)
Pethia nigripinna (Knight, Rema Devi, Indra & Arunachalam, 2012)	Freshwater	ENWG	4.5 cm SL	Fisheries: of no interest	NE	Bijukumar & Raghavan (2015)
Pethia pookodensis (Mercy & Jacob, 2007)	Freshwater	ENKL	4.3 cm SL	Fisheries: of no interest	CR	Bijukumar & Raghavan (2015)
Pethia punctata (Day, 1865)	Freshwater		7.5 cm	Fisheries: of no interest	LC	Easa & Basha (1995); Euphrasia (2004); Beevi & Ramachandran (2009); Bijukumar & Raghavan (2015); Plamoottil (2015)

Order, Family, Scientific name	Environment	Endemic, Exotic	Maximum size (TL)	Human Use	IUCN Status	References
Pethia ticto (Hamilton, 1822)	Freshwater		10.0 cm	Ornamental	LC	Easa & Basha (1995); Kurup *et al.* (2004); Raghavan *et al.* (2008); Johnson & Arunachalam (2009); Baby *et al.* (2010); Radhakrishnan & Kurup (2010); Abraham *et al.* (2011); Bijukumar *et al.* (2013); Bijukumar & Raghavan (2015); Plamoottil (2015)
Puntius amphibius (Valenciennes, 1842)	Freshwater; brackish		20.0 cm	Capture Fishery	DD	Easa & Basha (1995); Euphrasia (2004); Kurup *et al.* (2004); Raghavan *et al.* (2008); Beevi & Ramachandran (2009); Johnson & Arunachalam (2009); Baby *et al.* (2010); Radhakrishnan & Kurup (2010); Renjithkumar *et al.* (2011); Bijukumar *et al.* (2013); Plamoottil (2015)
Puntius arenatus (Day, 1878)	Freshwater	ENI	10.0 cm SL	Capture Fishery	VU	Johnson & Arunachalam (2009)
Puntius bimaculatus (Bleeker, 1863)	Freshwater		7.0 cm	Ornamental	LC	Euphrasia (2004); Raghavan *et al.* (2008); Johnson & Arunachalam (2009); Abraham *et al.* (2011); Bijukumar & Raghavan (2015)
Puntius burmanicus (Day, 1878)	Freshwater	EX	10.0 cm	Fisheries: of no interest	DD	Kurup *et al.* (2004)
Puntius cauveriensis (Hora, 1937)	Freshwater	ENCR	7.4 cm	Fisheries: of no interest	EN	Bijukumar & Raghavan (2015)
Puntius chola (Hamilton, 1822)	Freshwater		15.0 cm	Capture Fishery; Ornamental	LC	Easa & Basha (1995); Euphrasia (2004); Kurup *et al.* (2004); Beevi & Ramachandran (2009); Abraham *et al.* (2011); Bijukumar *et al.* (2013); Bijukumar & Raghavan (2015); Plamoottil (2015)
Pethia conchonius (Hamilton, 1822)	Freshwater		14.0 cm	Ornamental	LC	Easa & Basha (1995); Euphrasia (2004); Kurup *et al.* (2004); Johnson & Arunachalam (2009); Abraham *et al.* (2011); Bijukumar *et al.* (2013)

Order, Family, Scientific name	Environment	Endemic, Exotic	Maximum size (TL)	Human Use	IUCN Status	References
Puntius crescentus Yazdani & Singh, 1994	Freshwater	ENI	-	Ornamental	EN	Beevi & Ramachandran (2009)
Puntius dorsalis (Jerdon, 1849)	Freshwater		25.0 cm	Capture Fishery; Ornamental	LC	Euphrasia (2004); Kurup *et al.* (2004); Raghavan *et al.* (2008); Johnson & Arunachalam (2009); Abraham *et al.* (2011); Bijukumar & Raghavan (2015); Plamoottil (2015)
Puntius madhusoodani Kumar, Pereira & Radhakrishnan, 2012	Freshwater	ENKL	9.1 cm SL	Fisheries: of no interest	NE	Bijukumar & Raghavan (2015); Plamoottil (2015)
Puntius mahecola (Valenciennes, 1844)	Freshwater	ENKL	8.9 cm SL	Ornamental	DD	Raghavan *et al.* (2008); Abraham *et al.* (2011); Bijukumar *et al.* (2013); Bijukumar & Raghavan (2015); Plamoottil (2015)
Puntius melanostigma (Day, 1878)	Freshwater	ENI	10.0 cm	Fisheries: of no interest	NE	Easa & Basha (1995); Kurup *et al.* (2004); Beevi & Ramachandran (2009); Bijukumar & Raghavan (2015)
Puntius nelsoni Plamoottil, 2014	Freshwater	ENKL	9.1 cm SL	Ornamental	NE	Plamoottil (2015)
Puntius ophicephalus (Raj, 1941)	Freshwater	ENI	19.6 cm	Ornamental	EN	Arun (1999); Euphrasia (2004); Kurup *et al.* (2004); Radhakrishnan & Kurup (2010)
Puntius parrah Day, 1865	Freshwater	ENSI	15.0 cm	Capture Fishery	LC	Euphrasia (2004); Beevi & Ramachandran (2009); Johnson & Arunachalam (2009); Abraham *et al.* (2011); Bijukumar *et al.* (2013); Bijukumar & Raghavan (2015); Plamoottil (2015)
Puntius sophore (Hamilton, 1822)	Freshwater		20.0 cm	Ornamental	LC	Kurup *et al.* (2004); Johnson & Arunachalam (2009); Bijukumar *et al.* (2013); Bijukumar & Raghavan (2015)

Order, Family, Scientific name	Environment	Endemic, Exotic	Maximum size (TL)	Human Use	IUCN Status	References
Puntius thomassi (Day, 1874)	Freshwater	ENWG	100.0 cm	Capture Fishery	CR	Kurup *et al.* (2004); Beevi & Ramachandran (2009)
Puntius viridis Plamoottil & Abraham, 2014	Freshwater	ENKL	8.1 cm SL	Ornamental	NE	Plamoottil (2015)
Puntius vittatus Day, 1865	Freshwater		5.0 cm	Ornamental	LC	Easa & Basha (1995); Euphrasia (2004); Kurup *et al.* (2004); Raghavan *et al.* (2008); Beevi & Ramachandran (2009); Baby *et al.* (2010); Radhakrishnan & Kurup (2010); Abraham *et al.* (2011); Bijukumar *et al.* (2013); Bijukumar & Raghavan (2015); Plamoottil (2015)
Rasbora caverii (Jerdon, 1849)	Freshwater; brackish		10.0 cm	Capture Fishery; Ornamental	LC	Johnson & Arunachalam (2009)
Rasbora daniconius (Hamilton, 1822)	Freshwater; brackish		15.0 cm	Capture Fishery; Ornamental	LC	Easa & Basha (1995); Arun (1999); Euphrasia (2004); Kurup *et al.* (2004); Raghavan *et al.* (2008); Beevi & Ramachandran (2009); Baby *et al.* (2010); Radhakrishnan & Kurup (2010); Abraham *et al.* (2011); Bijukumar *et al.* (2013); Bijukumar & Raghavan (2015); Plamoottil (2015)
Sahyadria chalakkudiensis (Menon, Rema Devi & Thobias, 1999)	Freshwater	ENKL	12.5 cm	Ornamental	EN	Kurup *et al.* (2004); Bijukumar & Raghavan (2015)
Sahyadria denisonii (Day, 1865)	Freshwater	ENWG	15.0 cm	Ornamental	EN	Easa & Basha (1995); Euphrasia (2004); Kurup *et al.* (2004); Raghavan *et al.* (2008); Beevi & Ramachandran (2009); Johnson & Arunachalam (2009); Baby *et al.* (2010); Radhakrishnan & Kurup (2010); Abraham *et al.* (2011); Bijukumar *et al.* (2013); Bijukumar & Raghavan (2015)

Order, Family, Scientific name	Environment	Endemic, Exotic	Maximum size (TL)	Human Use	IUCN Status	References
Salmophasia acinaces (Valenciennes, 1844)	Freshwater	ENI	15.0 cm	Capture Fishery	LC	Easa & Basha (1995); Euphrasia (2004); Kurup *et al.* (2004); Beevi & Ramachandran (2009); Bijukumar & Raghavan (2015)
Salmophasia boopis (Day, 1874)	Freshwater	ENI	12.0 cm	Capture Fishery	LC	Easa & Basha (1995); Euphrasia (2004); Kurup *et al.* (2004); Beevi & Ramachandran (2009); Johnson & Arunachalam (2009); Baby *et al.* (2010); Radhakrishnan & Kurup (2010); Bijukumar *et al.* (2013); Bijukumar & Raghavan (2015); Plamoottil (2015)
Salmophasia balookee (Sykes, 1839)	Freshwater		15.0 cm	Capture Fishery	LC	Kurup *et al.* (2004); Johnson & Arunachalam (2009); Abraham *et al.* (2011); Bijukumar *et al.* (2013); Bijukumar & Raghavan (2015)
Salmophasia sardinella (Valenciennes, 1844)	Freshwater		15.0 cm	Capture Fishery	LC	Euphrasia (2004); Kurup *et al.* (2004)
Systomus chryseus Plamoottil, 2014	Freshwater	ENKL	15.2 cm SL	Ornamental	NE	Plamoottil (2015)
Systomus rufus Plamoottil, 2014	Freshwater	ENKL	10.6 cm SL	Ornamental	NE	Plamoottil (2015)
Systomus sarana (Hamilton, 1822)	Freshwater; brackish		42.0 cm	Capture Fishery; Gamefish; Ornamental	LC	Easa & Basha (1995); Euphrasia (2004); Kurup *et al.* (2004); Raghavan *et al.* (2008); Beevi & Ramachandran (2009); Johnson & Arunachalam (2009); Baby *et al.* (2010); Abraham *et al.* (2011); Renjithkumar *et al.* (2011); Bijukumar *et al.* (2013); Bijukumar & Raghavan (2015); Plamoottil (2015)
Tor khudree (Sykes, 1839)	Freshwater		50.0 cm	Capture Fishery; Aquaculture; Gamefish	EN	Easa & Basha (1995); Arun (1999); Euphrasia (2004); Kurup *et al.* (2004); Raghavan *et al.* (2008); Johnson & Arunachalam (2009); Radhakrishnan & Kurup (2010); Ambili *et al.* (2014); Bijukumar & Raghavan (2015)

Order, Family, Scientific name	Environment	Endemic, Exotic	Maximum size (TL)	Human Use	IUCN Status	References
Tor malabaricus (Jerdon, 1849)	Freshwater	ENI	40.0 cm	Gamefish; Ornamental	EN	Baby *et al.* (2010); Abraham *et al.* (2011); Bijukumar *et al.* (2013); Ambili *et al.* (2014); Bijukumar & Raghavan (2015)
Hypselobarbus mussullah (Sykes, 1839)	Freshwater	ENI	150 cm	Gamefish; Ornamental	EN	Kurup *et al.* (2004); Ambili *et al.* (2014); Bijukumar & Raghavan (2015)
Tor putitora (Hamilton, 1822)	Freshwater		275 cm	Capture Fishery; Aquaculture; Gamefish; Ornamental	EN	Euphrasia (2004); Kurup *et al.* (2004); Ambili *et al.* (2014)
Tor remadevii Kurup & Radhakrishnan, 2011	Freshwater	ENKL	33.182 cm	Capture Fishery	NE	Bijukumar & Raghavan (2015)
Tor tor (Hamilton, 1822)	Freshwater		200 cm	Capture Fishery; Aquaculture; Gamefish	NT	Kurup *et al.* (2004); Ambili *et al.* (2014)
Order: Cyprinodontiformes						
Family: Aplocheilidae						
Aplocheilus blockii Arnold, 1911	Freshwater; brackish		6.0 cm	Ornamental	LC	Euphrasia (2004); Kurup *et al.* (2004); Beevi & Ramachandran (2009); Abraham *et al.* (2011); Bijukumar *et al.* (2013); Bijukumar & Raghavan (2015)
Aplocheilus lineatus (Valenciennes, 1846)	Freshwater; brackish	ENPI	10.0 cm	Ornamental	LC	Easa & Basha (1995); Arun (1999); Euphrasia (2004); Kurup *et al.* (2004); Raghavan *et al.* (2008); Beevi & Ramachandran (2009); Johnson & Arunachalam (2009); Baby *et al.* (2010); Radhakrishnan & Kurup (2010); Abraham *et al.* (2011); Bijukumar *et al.* (2013); Bijukumar & Raghavan (2015); Plamoottil (2015)

Order, Family, Scientific name	Environment	Endemic, Exotic	Maximum size (TL)	Human Use	IUCN Status	References
Aplocheilus panchax (Hamilton, 1822)	Freshwater; brackish		9.0 cm	Capture Fishery; Ornamental	LC	Beevi & Ramachandran (2009); Johnson & Arunachalam (2009)
Family: Poeciliidae						
Gambusia affinis (Baird & Girard, 1853)	Freshwater; brackish	EX	4.0 cm	Capture Fishery; Ornamental	LC	Raghavan *et al.* (2008); Bijukumar & Raghavan (2015)
Poecilia reticulata Peters, 1859	Freshwater; brackish	EX	5.0 cm SL	Ornamental	NE	Easa & Basha (1995); Raghavan *et al.* (2008); Radhakrishnan & Kurup (2010); Bijukumar & Raghavan (2015)
Xiphophorus maculatus (Günther, 1866)	Freshwater	EX	4.0 cm	Ornamental	NE	Raghavan *et al.* (2008)
Order: Elopiformes						
Family: Elopidae						
Elops machnata (Forsskål, 1775)	Marine; freshwater; brackish		118 cm FL	Capture Fishery; Gamefish	LC	Raghavan *et al.* (2008); Bijukumar *et al.* (2013)
Family: Megalopidae						
Megalops cyprinoides (Broussonet, 1782)	Marine; freshwater; brackish		150 cm	Capture Fishery; Aquaculture; Gamefish	DD	Kurup *et al.* (2004); Raghavan *et al.* (2008); Bijukumar *et al.* (2013); Plamoottil (2015)
Order: Gonorynchiformes						
Family: Chanidae						
Chanos chanos (Forsskål, 1775)	Marine; freshwater; brackish		180 cm SL	Capture Fishery; Aquaculture; Gamefish	NE	Seethal *et al.* (2014)

Order, Family, Scientific name	Environment	Endemic, Exotic	Maximum size (TL)	Human Use	IUCN Status	References
Order: Mugiliformes						
Family: Mugilidae						
Mugil cephalus Linnaeus, 1758	Marine; freshwater; brackish		100.0 cm SL	Capture Fishery; Aquaculture; Gamefish	LC	Bijukumar *et al.* (2013)
Order: Osteoglossiformes						
Family: Notopteridae						
Notopterus notopterus (Pallas, 1769)	Freshwater; brackish		60.0 cm SL	Capture Fishery; Aquaculture; Ornamental	LC	Easa & Basha (1995); Euphrasia (2004); Kurup *et al.* (2004); Abraham *et al.* (2011); Bijukumar *et al.* (2013)
Order: Perciformes						
Family: Ambassidae						
Ambassis ambassis (Lacepède, 1802)	Marine; freshwater; brackish		15.0 cm	Ornamental	LC	Bijukumar *et al.* (2013); Plamoottil (2015)
Ambassis gymnocephalus (Lacepède, 1802)	Marine; freshwater; brackish		16.0 cm	Capture Fishery	LC	Euphrasia (2004); Kurup *et al.* (2004)
Ambassis nalua (Hamilton, 1822)	Marine; freshwater; brackish		12.5 cm	Capture Fishery	LC	Kurup *et al.* (2004)
Chanda nama Hamilton, 1822	Freshwater; brackish		11.0 cm	Capture Fishery; Ornamental	LC	Kurup *et al.* (2004); Beevi & Ramachandran (2009); Abraham *et al.* (2011)

Order, Family, Scientific name	Environment	Endemic, Exotic	Maximum size (TL)	Human Use	IUCN Status	References
Parambassis baculis (Hamilton, 1822)	Freshwater	ENI	5.0 cm SL	Fisheries: of no interest	LC	Johnson & Arunachalam (2009)
Parambassis dayi (Bleeker, 1874)	Freshwater; brackish	ENI	17.5 cm	Ornamental	LC	Euphrasia (2004); Kurup *et al.* (2004); Raghavan *et al.* (2008); Beevi & Ramachandran (2009); Radhakrishnan & Kurup (2010); Abraham *et al.* (2011); Renjithkumar *et al.* (2011); Bijukumar *et al.* (2013); Plamoottil (2015)
Parambassis ranga (Hamilton, 1822)	Freshwater; brackish		8.0 cm	Ornamental	LC	Easa & Basha (1995); Kurup *et al.* (2004); Johnson & Arunachalam (2009); Abraham *et al.* (2011)
Parambassis thomassi (Day, 1870)	Marine; freshwater; brackish	ENWG	12.0 cm SL	Ornamental	LC	Easa & Basha (1995); Euphrasia (2004); Kurup *et al.* (2004); Raghavan *et al.* (2008); Beevi & Ramachandran (2009); Radhakrishnan & Kurup (2010); Abraham *et al.* (2011); Renjithkumar *et al.* (2011); Bijukumar *et al.* (2013); Plamoottil (2015)
Family: Anabantidae						
Anabas testudineus (Bloch, 1792)	Freshwater; brackish		25.0 cm	Capture Fishery; Aquaculture; Ornamental	DD	Euphrasia (2004); Kurup *et al.* (2004); Raghavan *et al.* (2008); Beevi & Ramachandran (2009); Baby *et al.* (2010); Radhakrishnan & Kurup (2010); Abraham *et al.* (2011); Renjithkumar *et al.* (2011); Bijukumar *et al.* (2013); Bijukumar & Raghavan (2015); Plamoottil (2015)
Family: Channidae						
Channa diplogramma (Day, 1865)	Freshwater	ENKL	53.4 cm	Capture Fishery	VU	Abraham *et al.* (2011); Renjithkumar *et al.* (2011); Bijukumar & Raghavan (2015)

Order, Family, Scientific name	Environment	Endemic, Exotic	Maximum size (TL)	Human Use	IUCN Status	References
Channa gachua (Hamilton, 1822)	Freshwater		20.0 cm SL	Ornamental	LC	Easa & Basha (1995); Arun (1999); Kurup *et al.* (2004); Beevi & Ramachandran (2009); Radhakrishnan & Kurup (2010); Abraham *et al.* (2011); Bijukumar *et al.* (2013); Bijukumar & Raghavan (2015); Plamoottil (2015)
Channa marulius (Hamilton, 1822)	Freshwater		183 cm	Capture Fishery; Aquaculture; Gamefish; Ornamental	LC	Easa & Basha (1995); Euphrasia (2004); Kurup *et al.* (2004); Raghavan *et al.* (2008); Beevi & Ramachandran (2009); Baby *et al.* (2010); Radhakrishnan & Kurup (2010); Abraham *et al.* (2011); Renjithkumar *et al.* (2011); Bijukumar *et al.* (2013); Plamoottil (2015)
Channa micropeltes (Cuvier, 1831)	Freshwater		130 cm SL	Capture Fishery; Aquaculture; Gamefish; Ornamental	LC	Euphrasia (2004); Kurup *et al.* (2004); Plamoottil (2015)
Channa orientalis Bloch & Schneider, 1801	Freshwater		33.0 cm	Capture Fishery; Ornamental	NE	Easa & Basha (1995); Beevi & Ramachandran (2009); Johnson & Arunachalam (2009); Plamoottil (2015)
Channa punctata (Bloch, 1793)	Freshwater; brackish		31.0 cm	Capture Fishery; Aquaculture; Ornamental	LC	Euphrasia (2004); Kurup *et al.* (2004); Bijukumar & Raghavan (2015)
Channa striata (Bloch, 1793)	Freshwater; brackish		100.0 cm SL	Capture Fishery; Aquaculture; Ornamental	LC	Arun (1999); Easa & Basha (1995); Euphrasia (2004); Kurup *et al.* (2004); Raghavan *et al.* (2008); Beevi & Ramachandran (2009); Baby *et al.* (2010); Radhakrishnan & Kurup (2010); Abraham *et al.* (2011); Renjithkumar *et al.* (2011); Bijukumar *et al.* (2013); Bijukumar & Raghavan (2015); Plamoottil (2015)

Order, Family, Scientific name	Environment	Endemic, Exotic	Maximum size (TL)	Human Use	IUCN Status	References
Family: Cichlidae						
Etroplus maculatus (Bloch, 1795)	Freshwater; brackish		8.0 cm	Capture Fishery; Ornamental	LC	Easa & Basha (1995); Euphrasia (2004); Kurup *et al.* (2004); Raghavan *et al.* (2008); Beevi & Ramachandran (2009); Baby *et al.* (2010); Radhakrishnan & Kurup (2010); Abraham *et al.* (2011); Bijukumar *et al.* (2013); Plamoottil (2015)
Etroplus suratensis (Bloch, 1790)	Freshwater; brackish		40.0 cm	Capture Fishery; Aquaculture; Ornamental	LC	Easa & Basha (1995); Euphrasia (2004); Kurup *et al.* (2004); Raghavan *et al.* (2008); Beevi & Ramachandran (2009); Baby *et al.* (2010); Abraham *et al.* (2011); Renjithkumar *et al.* (2011); Bijukumar *et al.* (2013); Plamoottil (2015)
Oreochromis mossambicus (Peters, 1852)	Freshwater; brackish	EX	39.0 cm SL	Capture Fishery; Aquaculture; Gamefish; Ornamental	NT	Easa & Basha (1995); Arun (1999); Euphrasia (2004); Kurup *et al.* (2004); Raghavan *et al.* (2008); Radhakrishnan & Kurup (2010); Abraham *et al.* (2011); Bijukumar *et al.* (2013); Plamoottil (2015)
Oreochromis niloticus (Linnaeus, 1758)	Freshwater; brackish	EX	60.0 cm SL	Capture Fishery; Aquaculture	NE	Bijukumar *et al.* (2013)
Family: Eleotridae						
Eleotris fusca (Forster, 1801)	Freshwater; brackish		26.0 cm	Capture Fishery; Ornamental	LC	Euphrasia (2004); Kurup *et al.* (2004); Raghavan *et al.* (2008); Bijukumar *et al.* (2013)
Family: Gerreidae						
Gerres filamentosus (Cuvier, 1829)	Marine; freshwater; brackish		35.0 cm	Capture Fishery	LC	Plamoottil (2015)

Order, Family, Scientific name	Environment	Endemic, Exotic	Maximum size (TL)	Human Use	IUCN Status	References
Family: Gobiidae						
Awaous grammepomus (Bleeker, 1849)	Freshwater; brackish		15.0 cm SL	Capture Fishery; Ornamental	LC	Seethal *et al.* (2014)
Awaous stamineus (Eydoux & Souleyet, 1850)	Freshwater	EX	-	Ornamental	NE	Beevi & Ramachandran (2009)
Glossogobius giuris (Hamilton, 1822)	Marine; freshwater; brackish		50.0 cm SL	Capture Fishery; Aquaculture; Ornamental	LC	Easa & Basha (1995); Euphrasia (2004); Kurup *et al.* (2004); Raghavan *et al.* (2008); Beevi & Ramachandran (2009); Baby *et al.* (2010); Radhakrishnan & Kurup (2010); Abraham *et al.* (2011); Bijukumar *et al.* (2013); Plamoottil (2015)
Sicyopterus griseus (Day, 1877)	Freshwater; brackish	ENSI	9.7 cm	Ornamental	LC	Euphrasia (2004); Kurup *et al.* (2004); Raghavan *et al.* (2008); Baby *et al.* (2010); Abraham *et al.* (2011); Bijukumar *et al.* (2013)
Schismatogobius deraniyagalai Kottelat & Pethiyagoda, 1989	Freshwater		4.5 cm	Ornamental	DD	Easa & Basha (1995); Kurup *et al.* (2004)
Family: Latidae						
Lates calcarifer (Bloch, 1790)	Marine; freshwater; brackish		200 cm	Capture Fishery; Aquaculture; Gamefish; Ornamental	NE	Plamoottil (2015)
Family: Nandidae						
Nandus nandus (Hamilton, 1822)	Freshwater; brackish		20.0 cm	Capture Fishery; Ornamental	LC	Euphrasia (2004); Kurup *et al.* (2004); Raghavan *et al.* (2008); Beevi & Ramachandran (2009); Renjithkumar *et al.* (2011); Bijukumar *et al.* (2013); Plamoottil (2015)

Order, Family, Scientific name	Environment	Endemic, Exotic	Maximum size (TL)	Human Use	IUCN Status	References
Family: Osphronemidae						
Osphronemus goramy Lacepède, 1801	Freshwater; brackish	EX	70.0 cm SL	Capture Fishery; Aquaculture; Ornamental	LC	Raghavan *et al.* (2008); Bijukumar *et al.* (2013); Plamoottil (2015)
Pseudosphromenus cupanus (Cuvier, 1831)	Freshwater; brackish		7.5 cm	Ornamental	LC	Easa & Basha (1995); Euphrasia (2004); Kurup *et al.* (2004); Raghavan *et al.* (2008); Beevi & Ramachandran (2009); Abraham *et al.* (2011); Bijukumar *et al.* (2013); Bijukumar & Raghavan (2015); Plamoottil (2015)
Pseudosphromenus dayi (Köhler, 1908)	Freshwater; brackish	ENKL	7.5 cm	Ornamental	VU	Beevi & Ramachandran (2009); Bijukumar & Raghavan (2015)
Family: Pristolepididae						
Pristolepis fasciata (Bleeker, 1851)	Freshwater	ENWG	20.0 cm	Capture Fishery; Ornamental	LC	Kurup *et al.* (2004)
Pristolepis marginata Jerdon, 1849	Freshwater	ENKL	15.0 cm SL	Capture Fishery; Ornamental	LC	Easa & Basha (1995); Euphrasia (2004); Kurup *et al.* (2004); Raghavan *et al.* (2008); Baby *et al.* (2010); Radhakrishnan & Kurup (2010); Abraham *et al.* (2011); Renjithkumar *et al.* (2011); Bijukumar *et al.* (2013); Plamoottil (2015)
Pristolepis rubripinnis Britz, Kumar & Baby 2012	Freshwater	ENKL	12.9 cm SL	Ornamental	NE	Plamoottil (2015)
Family: Scatophagidae						
Scatophagus argus (Linnaeus, 1766)	Marine; freshwater; brackish		38.0 cm	Capture Fishery; Aquaculture; Ornamental	LC	Raghavan *et al.* (2008); Bijukumar *et al.* (2013); Plamoottil (2015)

Order, Family, Scientific name	Environment	Endemic, Exotic	Maximum size (TL)	Human Use	IUCN Status	References
Family: Terapontidae						
Terapon jarbua (Forsskål, 1775)	Marine; freshwater; brackish		36.0 cm	Capture Fishery	LC	Bijukumar *et al.* (2013); Seethal *et al.* (2014)
Order: Pleuronectiformes						
Family: Soleidae						
Brachirus orientalis (Bloch & Schneider, 1801)	Marine; freshwater; brackish		30.0 cm SL	Capture Fishery	NE	Easa & Basha (1995); Raghavan *et al.* (2008); Plamoottil (2015)
Order: Salmoniformes						
Family: Salmonidae						
Oncorhynchus mykiss (Walbaum, 1792)	Marine; freshwater; brackish	EX	122 cm	Capture Fishery; Aquaculture; Gamefish	NE	Kurup *et al.* (2004)
Order: Siluriformes						
Family: Bagridae						
Batasio travancoria Hora & Law, 1941	Freshwater	ENKL	10.0 cm	Fisheries: of no interest	VU	Easa & Basha (1995); Euphrasia (2004); Kurup *et al.* (2004); Raghavan *et al.* (2008); Johnson & Arunachalam (2009); Baby *et al.* (2010); Radhakrishnan & Kurup (2010); Abraham *et al.* (2011); Bijukumar *et al.* (2013); Bijukumar & Raghavan (2015); Plamoottil (2015)

Order, Family, Scientific name	Environment	Endemic, Exotic	Maximum size (TL)	Human Use	IUCN Status	References
Hemibagrus menoda (Hamilton, 1822)	Freshwater	ENI	45.0 cm	Capture Fishery	LC	Kurup *et al.* (2004)
Hemibagrus punctatus (Jerdon, 1849)	Freshwater	ENCR	45.0 cm	Capture Fishery	CR	Easa & Basha (1995); Bijukumar *et al.* (2013); Bijukumar & Raghavan (2015)
Horabagrus brachysoma (Günther, 1864)	Freshwater; brackish	ENKL	45.0 cm	Capture Fishery; Ornamental	VU	Euphrasia (2004); Kurup *et al.* (2004); Raghavan *et al.* (2008); Beevi & Ramachandran (2009); Baby *et al.* (2010); Abraham *et al.* (2011); Renjithkumar *et al.* (2011); Bijukumar & Raghavan (2015); Plamoottil (2015)
Horabagrus melanosoma Plamoottil & Abraham, 2013	Freshwater	ENKL	23.5 cm SL	Fisheries: of no interest	NE	Plamoottil (2015)
Horabagrus nigricollaris Pethiyagoda & Kottelat, 1994	Freshwater	ENKL	27.0 cm	Ornamental	EN	Euphrasia (2004); Kurup *et al.* (2004); Raghavan *et al.* (2008); Bijukumar & Raghavan (2015)
Mystus armatus (Day, 1865)	Freshwater; brackish	ENI	14.5 cm SL	Capture Fishery; Ornamental	LC	Easa & Basha (1995); Euphrasia (2004); Kurup *et al.* (2004); Raghavan *et al.* (2008); Beevi & Ramachandran (2009); Johnson & Arunachalam (2009); Baby *et al.* (2010); Radhakrishnan & Kurup (2010); Abraham *et al.* (2011); Renjithkumar *et al.* (2011); Bijukumar & Raghavan (2015)
Mystus bleekeri (Day, 1877)	Freshwater		15.5 cm	Capture Fishery; Ornamental	LC	Euphrasia (2004); Abraham *et al.* (2011)
Mystus canarensis Grant, 1999	Freshwater	ENI	10.1 cm SL	Fisheries: of no interest	NE	Plamoottil (2015)
Mystus cavasius (Hamilton, 1822)	Freshwater; brackish		40.0 cm SL	Capture Fishery	LC	Easa & Basha (1995); Euphrasia (2004); Kurup *et al.* (2004); Raghavan *et al.* (2008); Abraham *et al.* (2011); Bijukumar & Raghavan (2015)

Order, Family, Scientific name	Environment	Endemic, Exotic	Maximum size (TL)	Human Use	IUCN Status	References
Mystus gulio (Hamilton, 1822)	Freshwater; brackish		46.0 cm	Capture Fishery	LC	Euphrasia (2004); Kurup *et al.* (2004); Raghavan *et al.* (2008); Bijukumar *et al.* (2013); Plamoottil (2015)
Mystus heoki Plamoottil & Abraham, 2013	Freshwater	ENKL	13.7 cm SL	Fisheries: of no interest	NE	Plamoottil (2015)
Mystus indicus Plamoottil & Abraham, 2013	Freshwater	ENKL	10.7 cm SL	Fisheries: of no interest	NE	Plamoottil (2015)
Mystus keralai Plamoottil & Abraham, 2014	Freshwater	ENKL	5.9 cm SL	Fisheries: of no interest	NE	Plamoottil (2015)
Mystus keletius (Valenciennes, 1840)	Freshwater		18.0 cm	Capture Fishery; Ornamental	LC	Kurup *et al.* (2004); Raghavan *et al.* (2008); Abraham *et al.* (2011)
Mystus malabaricus (Jerdon, 1849)	Freshwater; brackish	ENWG	15.0 cm	Capture Fishery	NT	Easa & Basha (1995); Raghavan *et al.* (2008); Beevi & Ramachandran (2009); Baby *et al.* (2010); Abraham *et al.* (2011); Bijukumar *et al.* (2013); Bijukumar & Raghavan (2015); Plamoottil (2015)
Mystus menoni Plamoottil & Abraham, 2013	Freshwater	ENKL	12.1 cm SL	Fisheries: of no interest	NE	Plamoottil (2015)
Mystus montanus (Jerdon, 1849)	Freshwater; brackish	ENI	15.0 cm	Capture Fishery	LC	Easa & Basha (1995); Beevi & Ramachandran (2009); Abraham *et al.* (2011); Bijukumar *et al.* (2013); Bijukumar & Raghavan (2015)
Mystus oculatus (Valenciennes, 1840)	Freshwater; brackish	ENKL, ENTN	15.0 cm	Capture Fishery	LC	Easa & Basha (1995); Euphrasia (2004); Kurup *et al.* (2004); Raghavan *et al.* (2008); Abraham *et al.* (2011); Bijukumar *et al.* (2013); Bijukumar & Raghavan (2015); Plamoottil (2015)
Mystus seengtee (Sykes, 1839)	Freshwater		15.6 cm SL	Capture Fishery	LC	Bijukumar *et al.* (2013)

Order, Family, Scientific name	Environment	Endemic, Exotic	Maximum size (TL)	Human Use	IUCN Status	References
Mystus vittatus (Bloch, 1794)	Freshwater; brackish		21.0 cm SL	Capture Fishery; Ornamental	LC	Beevi & Ramachandran (2009); Abraham *et al.* (2011); Bijukumar & Raghavan (2015)
Sperata aor (Hamilton, 1822)	Freshwater		180 cm	Capture Fishery	LC	Kurup *et al.* (2004)
Sperata seenghala (Sykes, 1839)	Freshwater; brackish		150 cm	Capture Fishery; Aquaculture; Gamefish	LC	Bijukumar & Raghavan (2015)
Family: Clariidae						
Clarias batrachus (Linnaeus, 1758)	Freshwater; brackish		47.0 cm	Capture Fishery; Aquaculture; Ornamental	LC	Binoy (2010)
Clarias dayi Hora, 1936	Freshwater	ENKL, ENTN	17.5 cm	Capture Fishery	NE	Kurup *et al.* (2004); Binoy (2010); Bijukumar & Raghavan (2015)
Clarias dussumieri Valenciennes, 1840	Freshwater	ENWG	25.0 cm	Capture Fishery	NT	Easa & Basha (1995); Kurup *et al.* (2004); Euphrasia (2004); Johnson & Arunachalam (2009); Binoy (2010); Radhakrishnan & Kurup (2010); Abraham *et al.* (2011); Bijukumar *et al.* (2013); Bijukumar & Raghavan (2015); Plamoottil (2015)
Clarias gariepinus (Burchell, 1822)	Freshwater	EX	170 cm	Capture Fishery; Aquaculture; Gamefish	LC	Kurup *et al.* (2004); Binoy (2010); Radhakrishnan & Kurup (2010); Renjithkumar *et al.* (2011); Bijukumar & Raghavan (2015); Plamoottil (2015)
Horaglanis abdulkalami Babu, 2012	Freshwater	ENKL	3.8 cm	Fisheries: of no interest	NE	Bijukumar & Raghavan (2015)
Horaglanis alikunhii Babu & Nayar, 2004	Freshwater	ENKL	-	Fisheries: of no interest	DD	Bijukumar & Raghavan (2015)

Order, Family, Scientific name	Environment	Endemic, Exotic	Maximum size (TL)	Human Use	IUCN Status	References
Horaglanis krishnai Menon, 1950	Freshwater	ENKL	4.2 cm	Fisheries: of no interest	DD	Kurup *et al.* (2004); Bijukumar & Raghavan (2015)
Family: Erethistidae						
Pseudolaguvia austrina Radhakrishnan, Sureshkumar & Ng, 2011	Freshwater	ENWG	3.56 cm SL	Fisheries: of no interest	DD	Bijukumar *et al.* (2013); Bijukumar & Raghavan (2015)
Family: Heteropneustidae						
Heteropneustes fossilis (Bloch, 1794)	Freshwater; brackish		30.0 cm	Capture Fishery; Aquaculture; Ornamental	LC	Easa & Basha (1995); Arun (1999); Euphrasia (2004); Kurup *et al.* (2004); Raghavan *et al.* (2008); Beevi & Ramachandran (2009); Radhakrishnan & Kurup (2010); Abraham *et al.* (2011); Renjithkumar *et al.* (2011); Bijukumar *et al.* (2013); Bijukumar & Raghavan (2015); Plamoottil (2015)
Family: Kryptoglanidae						
Kryptoglanis shajii Vincent & Thomas, 2011	Freshwater	ENKL	5.9 cm SL	Fisheries: of no interest	NE	Bijukumar & Raghavan (2015)
Family: Loricariidae						
Hypostomus plecostomus (Linnaeus, 1758)	Freshwater	EX	50.0 cm SL	Capture Fishery; Ornamental	NE	Bijukumar & Raghavan (2015)
Family: Pangasiidae						
Pangasianodon hypophthalmus (Sauvage, 1878)	Freshwater	EX	130 cm SL	Capture Fishery; Aquaculture; Ornamental	EN	Plamoottil (2015)

Order, Family, Scientific name	Environment	Endemic, Exotic	Maximum size (TL)	Human Use	IUCN Status	References
Pangasius pangasius (Hamilton, 1822)	Freshwater; brackish		300 cm SL	Capture Fishery; Aquaculture; Gamefish	LC	Kurup *et al.* (2004); Bijukumar & Raghavan (2015)
Family: Schilbeidae						
Proeutropiichthys taakree (Sykes, 1839)	Freshwater	ENWG	42.0 cm	Capture Fishery	LC	Johnson & Arunachalam (2009)
Pseudeutropius mitchelli Günther, 1864	Freshwater	ENI	90.0 cm	Ornamental	EN	Euphrasia (2004); Kurup *et al.* (2004); Raghavan *et al.* (2008); Bijukumar *et al.* (2013); Bijukumar & Raghavan (2015); Plamoottil (2015)
Silonia childreni (Sykes, 1839)	Freshwater	ENWG	48.0 cm	Capture Fishery	EN	Kurup *et al.* (2004)
Family: Siluridae						
Ompok bimaculatus (Bloch, 1794)	Freshwater; brackish		45.0 cm SL	Capture Fishery; Aquaculture; Ornamental	NT	Easa & Basha (1995); Arun (1999); Euphrasia (2004); Kurup *et al.* (2004); Raghavan *et al.* (2008); Beevi & Ramachandran (2009); Baby *et al.* (2010); Radhakrishnan & Kurup (2010); Abraham *et al.* (2011); Renjithkumar *et al.* (2011); Bijukumar *et al.* (2013); Bijukumar & Raghavan (2015); Plamoottil (2015)
Ompok malabaricus (Valenciennes, 1840)	Freshwater	ENWG	51.0 cm SL	Aquaculture	LC	Euphrasia (2004); Kurup *et al.* (2004); Beevi & Ramachandran (2009); Abraham *et al.* (2011); Bijukumar *et al.* (2013); Bijukumar & Raghavan (2015); Plamoottil (2015)
Pterocryptis wynaadensis (Day, 1873)	Freshwater	ENWG	30.0 cm	Fisheries: of no interest	EN	Euphrasia (2004); Kurup *et al.* (2004); Johnson & Arunachalam (2009); Bijukumar & Raghavan (2015)

Order, Family, Scientific name	Environment	Endemic, Exotic	Maximum size (TL)	Human Use	IUCN Status	References
Wallago attu (Bloch & Schneider, 1801)	Freshwater; brackish		240 cm	Capture Fishery; Gamefish	NT	Easa & Basha (1995); Euphrasia (2004); Raghavan *et al.* (2008); Beevi & Ramachandran (2009); Abraham *et al.* (2011); Renjithkumar *et al.* (2011); Bijukumar & Raghavan (2015); Plamoottil (2015)
Family: Sisoridae						
Glyptothorax anamalaiensis Silas, 1952	Freshwater	ENWG	-	Fisheries: of no interest	EN	Euphrasia (2004); Kurup *et al.* (2004); Baby *et al.* (2010); Bijukumar *et al.* (2013); Bijukumar & Raghavan (2015)
Glyptothorax annandalei Hora, 1923	Freshwater	ENI	11.5 cm	Fisheries: of no interest	LC	Easa & Basha (1995); Kurup *et al.* (2004); Raghavan *et al.* (2008); Baby *et al.* (2010); Radhakrishnan & Kurup (2010); Abraham *et al.* (2011); Bijukumar *et al.* (2013); Bijukumar & Raghavan (2015)
Glyptothorax davissinghi Manimekalan & Das, 1998	Freshwater	ENKL	-	Fisheries: of no interest	EN	Kurup *et al.* (2004); Bijukumar & Raghavan (2015)
Glyptothorax elankadensis Plamoottil & Abraham, 2013	Freshwater	ENKL	11.5 cm SL	Fisheries: of no interest	NE	Bijukumar & Raghavan (2015); Plamoottil (2015)
Glyptothorax housei Herre, 1942	Freshwater	ENKL	10.0 cm	Ornamental	EN	Kurup *et al.* (2004); Bijukumar & Raghavan (2015)
Glyptothorax lonah (Sykes, 1839)	Freshwater	ENI	15.0 cm	Fisheries: of no interest	LC	Euphrasia (2004); Kurup *et al.* (2004)
Glyptothorax madraspatanus (Day, 1873)	Freshwater	ENWG, ENCR	11.5 cm	Fisheries: of no interest	EN	Easa & Basha (1995); Arun (1999); Euphrasia (2004); Johnson & Arunachalam (2009); Abraham *et al.* (2011); Bijukumar & Raghavan (2015)
Glyptothorax malabarensis Gopi, 2010	Freshwater	ENKL	5.5 cm SL	Ornamental	DD	Bijukumar & Raghavan (2015)

Order, Family, Scientific name	Environment	Endemic, Exotic	Maximum size (TL)	Human Use	IUCN Status	References
Order: Syngnathiformes						
Family: Syngnathidae						
Microphis cuncalus (Hamilton, 1822)	Freshwater; brackish		17.5 cm SL	Fisheries: of no interest	LC	Easa & Basha (1995); Euphrasia (2004); Kurup *et al.* (2004); Abraham *et al.* (2011); Bijukumar *et al.* (2013); Bijukumar & Raghavan (2015)
Order: Synbranchiformes						
Family: Mastacembelidae						
Macrognathus albus Plamoottil & Abraham, 2014	Freshwater; brackish	ENKL	21.1 cm SL	Ornamental	NE	Plamoottil (2015)
Macrognathus aral (Bloch & Schneider, 1801)	Freshwater; brackish		63.5 cm	Capture Fishery	LC	Kurup *et al.* (2004); Renjithkumar *et al.* (2011)
Macrognathus fasciatus Plamoottil & Abraham, 2014	Freshwater	ENKL	30.6 cm SL	Capture Fishery	NE	Plamoottil (2015)
Macrognathus guentheri (Day, 1865)	Freshwater	ENI	29.9 cm SL	Capture Fishery; Ornamental	LC	Euphrasia (2004); Kurup *et al.* (2004); Beevi & Ramachandran (2009); Baby *et al.* (2010); Abraham *et al.* (2011); Bijukumar *et al.* (2013); Plamoottil (2015)
Mastacembelus armatus (Lacepède, 1800)	Freshwater; brackish		90.0 cm	Capture Fishery; Ornamental	LC	Easa & Basha (1995); Arun (1999); Euphrasia (2004); Kurup *et al.* (2004); Raghavan *et al.* (2008); Beevi & Ramachandran (2009); Johnson & Arunachalam (2009); Baby *et al.* (2010); Radhakrishnan & Kurup (2010); Abraham *et al.* (2011); Renjithkumar *et al.* (2011); Bijukumar *et al.* (2013); Plamoottil (2015)

Order, Family, Scientific name	Environment	Endemic, Exotic	Maximum size (TL)	Human Use	IUCN Status	References
Mastacembelus oatesii Boulenger, 1893	Freshwater	EX	37.0 cm	Fisheries: of no interest	EN	Kurup *et al.* (2004)
Family: Synbranchidae						
Monopterus fossorius (Nayar, 1951)	Marine; freshwater; brackish	ENKL	23.0 cm	Fisheries: of no interest	EN	Abraham *et al.* (2011)
Ophisternon bengalense McClelland, 1844	Freshwater; brackish		100.0 cm	Fisheries: of no interest	LC	Beevi & Ramachandran (2009)
Order: Tetraodontiformes						
Family: Tetraodontidae						
Carinotetraodon travancoricus (Hora & Nair, 1941)	Freshwater	ENKL	3.5 cm	Ornamental	VU	Easa & Basha (1995); Euphrasia (2004); Kurup *et al.* (2004); Raghavan *et al.* (2008); Beevi & Ramachandran (2009); Abraham *et al.* (2011); Bijukumar *et al.* (2013); Plamoottil (2015)

ENI = endemic to India; ENPI = endemic to peninsular India; ENSI = endemic to south India; ENWG = endemic to Western Ghats of India; ENCR = endemic to Cauvery River; ENKL = endemic to Kerala; ENTN = endemic to Tamil Nadu; EX – Exotic; TL = Total Length; SL = Standard Length; IUCN = International Union For Conservation of Nature; Vulnerable; NT – Near Threatened; LC – Least Concern; DD – Data Deficient; NE – Not Evaluated

Table 2. Number of family, genera and species under each Order of Freshwater Fishes of Kerala

Order	Family	Number of Genera	Number of Genera
Anguilliformes	Anguillidae	1	2
	Ophichthidae	1	1
Beloniformes	Belonidae	2	2
	Hemiramphidae	1	2
Clupeiformes	Clupeidae	1	1
Cypriniformes	Balitoridae	4	11
	Cobitidae	3	3
	Nemacheilidae	5	19
	Cyprinidae	32	121
Salmoniformes	Salmonidae	1	1
Cyprinodontiformes	Aplocheilidae	1	3
	Poeciliidae	3	3
Elopiformes	Elopidae	1	1
	Megalopidae	1	1
Gonorynchiformes	Chanidae	1	1
Mugiliformes	Mugilidae	1	1
Osteoglossiformes	Notopteridae	1	1
Perciformes	Ambassidae	3	8
	Anabantidae	1	1
	Channidae	1	7
	Cichlidae	2	4
	Eleotridae	1	1
	Gerreidae	1	1
	Gobiidae	4	5
	Latidae	1	1

Order	Family	Number of Genera	Number of Genera
	Nandidae	1	1
	Osphronemidae	2	3
	Pristolepididae	1	3
	Scatophagidae	1	1
	Terapontidae	1	1
Pleuronectiformes	Soleidae	1	1
Siluriformes	Bagridae	4	23
	Clariidae	2	7
	Erethistidae	1	1
	Heteropneustidae	1	1
	Kryptoglanidae	1	1
	Loricariidae	1	1
	Pangasiidae	2	2
	Schilbeidae	3	3
	Siluridae	3	4
	Sisoridae	1	8
Syngnathiformes	Syngnathidae	1	1
Synbranchiformes	Mastacembelidae	2	6
	Synbranchidae	2	2
Tetraodontiformes	Tetraodontidae	1	1
Total = 16	**Total** = 45	**Total** = 106	**Total** = 273

4

FRESHWATER FISH DIVERSITY OF TAMIL NADU

H. S. Mogalekar and P. Jawahar

Department of Fisheries Biology and Resource Management, Fisheries College and Research Institute, Tamil Nadu Fisheries University, Thoothukudi-628 008, Tamil Nadu, India

INTRODUCTION

Tamil Nadu lies in the southeast of the Indian Peninsula and is bounded by the Eastern Ghats on the north, by the Western Ghats (Nilgiri and Anamalai Hills) on the west, by the Bay of Bengal in the east and by the Indian Ocean on the south. The Western Ghats region starts from Gudalur taluk in the Nilgiris District and ends at Agatheeswaram taluk in Kanyakumari district with three important rivers Cauvery, Vaigai and Tambiraparani traveling through this region (Government of Tamilnadu, 2015). The Eastern Ghats, unlike the Western Ghats, are not a continuous range of mountains but a series of broken and weathered relicts of the Peninsular Plateau represented as a series of isolated hills. In the south, the Eastern Ghats merge with the Western Ghats and the eastern parts of the Niligiri, Anamalai and Palani Hills also form parts of the Eastern Ghats (Devi & Indra, 2003).

Tamil Nadu is enriched with a vast expanse of freshwater fisheries resources [3.08 lakh hectare (ha)] in the form of streams, rivers, canals (7400 kilometer), major reservoirs (52,000 ha), Big / small Irrigation tanks (98000 ha), small lakes and Rural Fishery Demonstration tanks (158000 ha) and cold water lakes and reservoirs (Government of Tamilnadu, 2015). Rivulets and rivers originating from Western Ghats and Eastern Ghats has been conferred with diverse fish germplasm resources (Devi and Indra, 2003; Johnson and Arunachalam, 2009; Raghavan *et al.*, 2013), has become the geographic focus of attention for ichthyologists and conservation biologists. 287 species of freshwater fishes have been reported from the Western Ghats of India (Shaji *et al.*, 2000). The fish diversity of Eastern Ghats has not been thoroughly studied as those of the Western Ghats (Devi and Indra, 2003). Devi and Indra (2003) listed 127 fish species under 26 families and nine orders from Eastern Ghats of India.

The comprehensive studies on the freshwater fishes from various districts of Tamil Nadu were conducted by Raj (1916), Jayaram *et al.* (1982), Indra (1991, 1992, 1993, 1994), Talwar and Jhingran (1991), Menon (1993), Devi and Raghunathan (1997), Devi *et al.* (1997), Devi *et al.* (1998), Devi and Indra (2000) Rajagopal *et al.* (2013), Ramanujam *et al.* (2014) and Mogalekar and Jawahar (2015). Since the publication by Devi and Indra (2000) on Freshwater ichthyofaunal resources of Tamil Nadu, numerous changes in taxonomy and / or systematic placement of species took place and several species of freshwater fishes in Tamil Nadu has been recorded. Information on species diversity of fishes reveals crustal facts necessary for the management, conservation and sustainable utilization of fishery resources. Restructured information on species composition of freshwater fishes from various localities of Tamil Nadu is document in this paper.

MATERIALS AND METHODS

The present list of the freshwater fish fauna from different localities of Tamil Nadu was compiled based on publications by Sreenivasan (1976); Talwar and Jhingran (1991); Devi *et al.* (1998); Molur and Walker (1998); Sreenivasan (1998); Arunachalam and Sankaranarayanan (1999); Devi and Raghunathan (1999); Arunachalam and Manimekalan (2000); Gopalakrishnan and Ponniah (2000); Devi and Indra (2000); Shaji *et al.* (2000); Jeyaraj (2000); Balasundaram *et al.* (2001); Knight and Devi (2009); Johnson and Arunachalam (2009); Jayaram (2010); Knight and Devi (2010); Dhinakaran *et al.* (2011); Knight *et al.* (2012); Ng (2013); Rajagopal *et al.* (2013); Knight and Devi (2014), Ramanujam *et al.* (2014) and Mogalekar and Jawahar (2015).

While listing the species of Tamil Nadu, catalog of fishes by Eschmeyer have been followed (Eschmeyer *et al.*, 2016). Details on the environment, endemic or exotic status and category (food or ornamental) of all species were obtained by retrieving species level information from the published literature as well as FishBase (Froese and Pauly, 2013). The endemicity or exoticity of all species indicated based on Indian subcontinent. The listing of food fishes is based on the maximum size of the species, acceptance by local people and discussion with fishermen. The list of ornamental fishes is prepared based on colouration pattern, shape and maximum size (Table 1). Information on the conservation status of all taxa in this paper was retrieved from the International Union for Conservation of Nature (IUCN) Red List of Threatened Species, whose underlying assessments are based on the IUCN Red List Categories and criteria (Version 2015.2) downloaded on 05 October 2015 (IUCN, 2015).

RESULTS AND DISCUSSION

Attempts have been made to include all the species of freshwater fishes known to occur in the streams, rivers, canals, reservoirs, Irrigation tanks and cold water lakes of Tamil Nadu. This paper discusses on the environment in which species occurs, endemicity or exoticity of species, present utilization of the fish resources (food or ornamental) and conservation status.

In total 218 species (including 125 were primary freshwater fishes, 68 were freshwater cum brackish and remaining 25 were freshwater cum brackish cum marine and 23 exotic species) so far known from the freshwater bodies of Tamil Nadu belonging to 95 generas, 35 families and 14 orders (Table 1 & 2). Cyprinidae contributed 48.16 % to total fish fauna with 105 species belonging to 32 genera, followed by Bagridae (13 species from three genera), Gobiidae (10 species from nine genera), Cichlidae (seven species from four genera), Ambassidae (seven species from three genera) and Nemacheilidae (seven species from two genera) (Table 2 & Figure 1). Many species could be distributed in the drainage of the Western Ghats and Eastern Ghats and other unexplored areas and therefore further biodiversity explorations are required. Cyprinidae with 32 genera and 105 species is the most speciose group among the freshwater fishes and many species form commercially important fisheries and most of interest to aquaculture (Table 2).

Talwar and Jhingran (1991) listed the 91 species of freshwater fishes from Tamil Nadu belonging to 48 generas; similarly Jayaram (2010) reported 99 species belonging to 51 genera. However, Devi and Indra (2000) described 144 species of freshwater fishes including nine exotic and six saltwater dispersents belonging to 63 genera, 27 families and 11 orders from Tamil Nadu. However, present record of 218 species is relatively higher compared to all the above reports. Since the publication of Devi and Indra (2000), 74 species has been reported from freshwater bodies in Tamil Nadu.

Indra (1992) recorded 42 fish species belonging to 15 families and 7 orders from Kanyakumari District of Tamil Nadu. Devi and Ilango (1993) reported 20 species belonging to 13 genera, seven families and five orders from Pudukkottai District of Tamil Nadu. Balasundaram *et al.* (1999) documented 24 species belonging to 18 genera and 13 families from Grand Anicut, Cauvery River in Tiruchirapalli District of Tamil Nadu. Devi *et al.* (1999) documented 51 species of primary freshwater fishes under 16 families and 7 orders from Chennai, Chengleput and Thiruvallur districts of Tamil Nadu. Shaji *et al.* (2000) described 41 species belonging to 27 genera from the Tamil Nadu region of Western Ghats. Balasundaram *et al.* (2001) identified 19 species of fish belonging to 15 genera and nine families from Kolli Hills, Western Ghats, Salem District of Tamil Nadu. Daniels and Rajagopal (2004) noted 32 species representing 21 genera and 13 families from Chembaramoakkam Lake in the outskirts of Chennai. Devi *et al.* (2007) described 60 species under 18 families and 8 orders from Indira Gandhi Wildlife Sanctuary. 22 species of freshwater fishes are known from streams of Tamil Nadu region of Western Ghats (Johnson and Arunachalam, 2009). 83 species of fishes belonging to 49 genera and 23 families are known to inhabit the fresh water habitats of Chennai (Knight and Devi, 2010). Rajagopal and Davidar (2013) recorded 10 species of catfishes representing four families from the 21 wetland of Kancheepuram and Kanyakumari Districts of Tamil Nadu. Ramanujam *et al.* (2014) noted 52 species of freshwater fishes from Adyar Wetland complex in Chennai. Mogalekar and Jawahar (2015) listed 156 species of freshwater ornamental fishes belonging to 68 generas, 27 families and eight orders from Tamil Nadu. All the above reports agreed with dominance of cyprinids over other freshwater fish families. The available literature pointed out the indigenous cyprinids, typically mahseer, *Puntius dubius* and *Cirrhinus cirrhosa* has completely disappeared due to introduced fishes like Catla, Rohu, Mrigal, *L. calbasu,* Common carp, Grass carp and Tilapia. These have been established themselves in most of the reservoirs by natural breeding. Barckishwater forms migrating upstream rivers (Example: *Hilsa*) and freshwater form migrating downstream (Example: *Anguilla*) has been nearly wiped out due to daming and construction of canals on rivers.

Barckishwater forms reported from the rivers and reservoirs are *Hilsa kelee* (Cuvier, 1829); *Tenualosa ilisha* (Hamilton, 1822); *Chelon parsia* (Hamilton, 1822); *Chelon planiceps* (Valenciennes, 1836); *Chelon subviridis* (Valenciennes, 1836); *Mugil cephalus* (Linnaeus, 1758); *Megalops cyprinoides* (Broussonet, 1782); *Chanos chanos* (Forsskål, 1775); *Amblyeleotris gymnocephala* (Bleeker, 1853); *Apocryptodon madurensis* (Bleeker, 1849); *Oxuderces dentatus* (Eydoux & Souleyet, 1850) and *Sicyopterus griseus* (Day, 1877) (Table 1). The present literature survey revealed presence of 93 saltwater dispersents visiting freshwater in Tamil Nadu, whereas Devi and Indra (2000) recorded 6 species saltwater dispersents from the records of Zoological Survey of India.

Of these, 79 species are endemic (36.23 % endemicity) to Indian subcontinent. 24 species are endemic to India, three species are endemic to peninsular India, three species are endemic to south India, 20 species are endemic to Western Ghats of India, 10 species are endemic to Cauvery River, four species are endemic to Kerala and 15 species are endemic to Tamil Nadu (Table 1). Species endemic to Tamil Nadu are *Chela macrolepis* Knight & Devi, 2014; *Dawkinsia tambraparniei* (Silas, 1954); *Garra kalakadensis* Devi, 1993; *Horalabiosa joshuai* Silas, 1954; *Horalabiosa palaniensis* Devi & Menon, 1994; *Hypselobarbus dubius* (Day, 1867); *Hypselobarbus kurali* Menon & Devi, 1995; *Osteobrama neilli* (Day, 1873); *Pethia sharmai* (Menon & Devi, 1993); *Puntius mudumalaiensis* Menon & Devi, 1992; *Nemacheilus nilgiriensis* Menon, 1987; *Nemacheilus triangularis* Day, 1865; *Mystus oculatus* (Valenciennes, 1840); *Clarias dayi* Hora, 1936 and *Heteropneustes longipectoralis* Devi & Raghunathan, 1999.

The exotic species in Tamil Nadu were nine in number till the publication of Devi and Indra (2000), however it has increased to 23 which includes, *Ctenopharyngodon idella* (Valenciennes, 1844); *Cyprinus carpio var. communis* (Linnaeus, 1758); *Cyprinus carpio var. specularis* (Lacepède, 1803); *Hypophthalmichthys molitrix* (Valenciennes, 1844); *Labeo altivelis* Peters, 1852; *Systomus rubripinnis* (Valenciennes 1842); *Tinca tinca* (Linnaeus, 1758); *Gambusia affinis* (Baird & Girard, 1853); *Poecilia reticulata* Peters, 1859; *Xiphophorus hellerii* Heckel, 1848; *Xiphophorus maculatus* (Günther, 1866); *Cichlasoma trimaculatum* (Gunther 1867); *Hemichromis bimaculatus* (Gill 1862); *Oreochromis aureus* (Steindachner 1864); *Oreochromis mossambicus* (Peters, 1852); *Oreochromis niloticus* (Linnaeus, 1758); *Osphronemus gourami* (Lacepède 1801); *Trichopodus trichopterus* (Pallas, 1770); *Oncorhynchus mykiss* (Walbaum, 1792); *Oncorhynchus nerka* (Walbaum, 1792); *Clarias gariepinus* (Burchell 1822); *Pterygoplichthys disjunctivus* (Weber, 1991) and *Pterygoplichthys pardalis* (Castelnau, 1855) (Table 1).

A consolidated list of freshwater fishes comprises of 139 species with ornamental value, 114 species with food value and 35 species with ornamental as well as food value in Tamil Nadu (Table 1). As numbers of ornamental as well as food fishes are high, seed production by selective breeding is recommended for income generation through culture of these species. Selective breeding and ranching of native fish species may help to overcome the problems of species endangerment.

Based on the information collected on conservation status of all taxa, four species listed as Critically Endangered, 20 as Endangered, 7 as Vulnerable, 10 as Near Threatened, one species as Lower Risk- Near Threatened, 135 as Least Concern, 11 as Data Deficient and 30 species have not yet been evaluated for their conservation status by IUCN Red list (Table 1). The overall aim of IUCN Red List is to convey urgency and scale of conservation problem to public and policy makers, and to motivate the global community to try to reduce species extinction.

As per IUCN conservation status, 24 endemic species are at great risk of endangerment. Identification, listing and prioritization of threats are required to conserve the endangered species of freshwater fishes in Tamil Nadu. Species listed as Critically Endangered *Barbodes bovanicus* (Day, 1877); *Neolissochilus wynaadensis* (Day, 1873); *Puntius thomassi* (Day, 1874) and *Hemibagrus punctatus* (Jerdon, 1849). The endangered category includes *Dawkinsia arulius* (Jerdon, 1849); *Dawkinsia tambraparniei* (Silas, 1954); *Devario neilgherriensis* (Day, 1867); *Garra hughi* (Silas, 1955); *Garra kalakadensis* (Devi, 1993); *Horalabiosa joshuai* Silas, 1954; *Hypselobarbus dubius* (Day, 1867); *Hypselobarbus curmuca* (Hamilton, 1807); *Hypselobarbus micropogon* (Valenciennes, 1842); *Labeo potail* (Sykes, 1839); *Pethia sharmai* (Menon & Devi, 1993); *Puntius cauveriensis* (Hora, 1937); *Tor khudree* (Sykes, 1839); *Tor malabaricus* (Jerdon, 1849); *Tor putitora* (Hamilton, 1822); *Nemacheilus pulchellus* (Day, 1873); *Clarias magur* (Hamilton 1822); *Pterocryptis wynaadensis* (Day, 1873); *Glyptothorax housei* (Herre, 1942) and *Glyptothorax madraspatanum* (Day, 1873). Species under Near Threatened category are *Anguilla bengalensis* (Gray, 1831); *Anguilla bicolor* (McClelland, 1844); *Hypophthalmichthys molitrix* (Valenciennes, 1844); *Labeo pangusia* (Hamilton, 1822); *Tor tor* (Hamilton, 1822); *Parambassis lala* (Hamilton 1822); *Mystus malabaricus* (Jerdon, 1849); *Clarias dussumieri* (Valenciennes, 1840); *Ompok bimaculatus* (Bloch, 1794) and *Wallago attu* (Bloch & Schneider, 1801) (Table 1). The species, *Laubuka fasciata* (Silas, 1958); *Cirrhinus cirrhosus* (Bloch, 1795); *Cyprinus carpio var. communis* (Linnaeus, 1758); *Horalabiosa palaniensis* (Devi & Menon, 1994); *Puntius arenatus* (Day, 1878); *Puntius mudumalaiensis* (Menon & Devi, 1992) and *Nemacheilus kodaguensis* (Menon, 1987) are Vulnerable.

CONCLUSION

Tamil Nadu is endowed with rich freshwater fish diversity. Proper conservation strategies need to be proposed, as 11.0 % of endemic species facing the threat of endangerment. Exotic as well as transplanted species are well established in 6o adopt. Daming and channelization across the rivers and streams affected on upstream and downstream migration of migratory species, proper migrating passages requisite to provided.

REFERENCES

Arunachalam, M. & Manimekalan, A. 2000. Economically important and cultivable fishes of the Nilgiri biosphere reserve. p. 237-239. In A.G. Ponniah and A. Gopalakrishnan (eds.) Endemic fish diversity of Western Ghats. NBFGR-NATP Publication. National Bureau of Fish Genetic Resources, Lucknow, U.P., India. 347 pp.

Arunachalam, M. & Sankaranarayanan, A. 1999. New records of fishes from Gadana river, south Tamil Nadu, India. Journal of the Bombay Natural History Society 96: 336-337.

Balasundaram, C., Dheepa, A. & Marippan, P. 1999. Fish Diversity in Grand Anicut, Cauvery River (Tiruchirapalli District, Tamil Nadu). Zoos' Print Journal 14 (8): 87-88.

Balasundaram, C., Arumugam, R. & Murugan, P.B. 2001. Fish Diversity of Kolli Hills, Western Ghats, Salem District, Tamil Nadu. Zoos' Print Journal 16(1): 403-406.

Chatterjee, T.K., Barman, R.P. & Mishra, S.S. 2013. Mangrove Associate Gobles (Teleostei: Gobioidei) of Indian Sundarbans. Records of the Zoological Survey of India 113(3): 59-77.

Daniels, R.J.R. & Rajagopal, B. 2004. Fishes of Chembarampakkam Lake - a wetland in the outskirts of Chennai. Zoos' Print Journal 19: 1481-1483; http://dx.doi.org/10.11609/JoTT.ZPJ.1041.1481-3

Devi, K.R. & Raghunathan, M.B. 1999. Heteropneustes longipectoralis (Siluriformes: Heteropneustidae) a new species from the Anamalai Hills, in the Western Ghats. Records of the Zoological Survey of India 97(3): 109-115.

Devi, K.R. & Indra, T.J. 2000. Freshwater ichthyofaunal resources of Tamil Nadu. pp. 77-97. In: Ponniah, A.G. and Gopalakrishnan, A. (eds.), Endemic Fish Diversity of Western Ghats. NBFGR – NATP Publication – 1,347 p. National Bureau of Fish Genetic Resources, Lucknow, U.P., India.

Devi, K.R. & Ilango, K. 1993. On a collection of fish from Pudukkottai, District Tamil Nadu. Records of the Zoological Survey of India 93 (1-2): 241 -251.

Devi, K.R. & Raghunathan, M.B. 1997. The Ichthyofauna of Dharmapuri District, Tamilnadu. Records of the Zoological Survey of India 97 (Part-I) : 145-162.

Devi, K.R. & Indra, T.J. 2003. An updated checklist of Ichthyofauna of Eastern Ghats. Zoos' Print 18 (4): 1067-1070; http://dx.doi.org/10.11609/JoTT.ZPJ.18.4.1067-70

Devi, K.R., Indra, T.J. & Raghunathan, M.B. 2007. Ichthyofauna of Indira Gandhi Wildlife Sanctuary, Tamil Nadu. Records of the Zoological Survey of India. Miscellaneous Publication Occasional Paper No. 277: 1-42 + 16 colour plates.

Devi, K.R., Indra, T.J., Raghunathan, M.B., Bai, M.M. & Ravichandran, M.S. 1997. Ichthyofauna of Tambraparni River system, Tamilnadu. Zoo's Print 12(7): 1-2.

Devi, K.R., Indra, T.J., Raghunathan, M.B. & Bai, M.M. 1999. On a collection of fish fauna from Chennai, Chengleput and Thiruvallur districts of Tamil Nadu. Records of the Zoological Survey of India 97(4): 151-166. http://faunaofindia.nic.in/PDFVolumes/records/097/04/0151-0166.pdf

Dhinakaran, A., Alikunhi, N.M., Chinnathambi, S., Sornam, R., Kalaiselvam, M., Rajasekaran, R. & Manivannan, S. 2011. Assessment of Morphometric and Genetic Variation in Three Freshwater Fish Species of the Genus Garra (Osteichthyes: Cyprinidae). Notulae Scientia Biologicae 3(1):12-16.

Eschmeyer, W. N., Fricke, R. and van der Laan, R. (eds.) 2016. Catalog of fishes: genera, species, references. Electronic version accessed 19-28 March 2016.

(http://researcharchive.calacademy.org/research/ichthyology/catalog/fishcatmain.asp). [This version was edited by Bill Eschmeyer.]

Froese, R. & Pauly, D. (Eds.) 2013. FishBase. World Wide Web electronic publication. www.fishbase.org, version (02/2013). 05 October 2015.

Gopalakrishnan, A & Ponniah, A.G. 2000. Cultivable, ornamental, sport and food fishes endemic to peninsular India with special reference to Western Ghats. pp. 77-97. In: Ponniah, A.G. and Gopalakrishnan, A. (eds.), Endemic Fish Diversity of Western Ghats. NBFGR – NATP Publication – 1,347 p. National Bureau of Fish Genetic Resources, Lucknow, U.P., India.

Government of Tamilnadu, 2015. Fisheries Department, Inland Fisheries. Retrieved 25 October 2015, from http://www.fisheries.tn.gov.in/Inland-main.html.

Indra, T.J. 1992. Report on the Ichthyofauna of Kanyakumari district, Tamilnadu. Records of the Zoological Survey of India 92(1-4): 177-192. http://faunaofindia.nic.in/PDFVolumes/records/092/01-04/0177-0192.pdf

Indra, T.J. 1994. On a collection of fishes from Tanjavur and Trichy districts, Tamilnadu. Records of the Zoological Survey of India 94(2-4): 403-433.

Indra, T.J. 1991. Report on the ichthyofauna of Anna and Madurai districts, Tamilnadu. Records of the Zoological Survey of India 89(1-4): 233-244.

Indra, T.J. 1993. Report on the Ichthyofauna from the mapping survey of Kamarajar and Pasumpon Ramalingam Districts. Records of the Zoological Survey of India 90(1-4): 61-68.

IUCN, 2015. The IUCN Red List of Threatened Species. Version2015.2. IUCN, Gland, Switzerland and Cambridge, UK, http://www.iucnredlist.org. 05 October 2015.

Jeyaraj, C.T. 2000. Biodiversity of fish species in the lotic habitats of Kanyakumari district. p. 177-179. In: Ponniah, A.G. and Gopalakrishnan, A. (eds.), Endemic Fish Diversity of Western Ghats. NBFGR – NATP Publication – 1,347 p. National Bureau of Fish Genetic Resources, Lucknow, U.P., India.

Jayaram, K.C. 2010. The Freshwater Fishes of the Indian Region. 2nd Edition, Narendra Publishing House, New Delhi, 616pp.

Jayaram, K.C., Venkateswarlu, T. & Raghunathan, M.B. 1982. A survey of the Cauvery River system with a major account of its fish fauna. Records of the Zoological Survey of India. Miscellaneous Publication Occasional Paper No. 36. 115p.

Johnson, J.A. & Arunachalam, M. 2009. Diversity, distribution and assemblage structure of fishes in streams of southern Western Ghats, India. Journal of Threatened Taxa 1(10): 507-513. http://threatenedtaxa.org/ZooPrintJournal/2009/October/o214626x09507-513.pdf

Knight, J.D.M. & Devi, K.R. 2009. On a record of Badis badis (Hamilton) (Teleostei: Perciformes: Badidae) from Tamil Nadu. Journal of the Bombay Natural History Society 106(2): 329-330.

Knight, J.D.M. & Devi, K.R. 2014. Chela macrolepis, a new species of cyprinid fishfrom southern India (Teleostei: Cyprinidae). Ichthyological Exploration of Freshwaters 25(2), 159-166.

Knight, J.D.M. & Devi, K.R. 2010. Species persistence: a re-look at the freshwater fish fauna of Chennai, India. Journal of Threatened Taxa 2(12): 1334-1337; http://dx.doi.org/10.11609/JoTT.o2519.1334-7

Knight, J.D.M. 2010. On a record of Puntius gelius (Hamilton, 1822) (Teleostei: Cypriniformes: Cyprinidae) from Tamil Nadu. Journal of Threatened Taxa 2(3): 786-787. http://threatenedtaxa.org/ZooPrintJournal/2010/March/o229826iii10786-787.pdf

Knight, J.D.M., Devi, K.R., Indra, T.J. & Arunachalam, M. 2012. A new species of barb Puntius nigripinnis (Teleostei: Cyprinidae) from southern Western Ghats, India. Journal of Threatened Taxa 4(3): 2409-2416. http://threatenedtaxa.org/ZooPrintJournal/2012/March/o301426iii122409-2416.pdf

Menon, A.G.K. 1993. Check List - Freshwater Fishes of India. Report submitted to Ministry of Environ. & Forests, Goverment of India, New Delhi 137p.

Mogalekar, H.S. & Jawahar, P. 2015. Freshwater ornamental fish diversity of Tamil Nadu. Journal of the Inland Fisheries Society of India 47(2): 27-37.

Molur, S. & Walker, S. 1998. Freshwater fishes of India. Conservation Assessment and Management Plan (CAMP) Workshop. Zoo Outreach Organisation, Tamil Nadu, India.

Ng, H.H. 2013. Ompok karunkodu, a new catfish (Teleostei: Siluridae) from southern India. Zootaxa 3694(2): 161-166.

Raghunathan, M.B. 1978. Studies on seasonal tanks in Tamilnadu. 1. Chambarambakkam tank. The India Journal of Zootomy 19(2): 81-85.

Raj, S.B. 1916. Notes on the freshwater fish of Madras. Records of the Indian Museum XII (Part VI): 249-294.

Rajagopal, B. & Davidar, P. 2013. Distribution of catfishes in wetlands of two flood plain districts in Tamil Nadu, India. Journal of Threatened Taxa 5(17): 5277–5282; http://dx.doi.org/10.11609/JoTT.o2889.5277-82

Ramanujam, M.E., Devi, K.R. & Indra, T.J. 2014. Ichthyofaunal diversity of the Adyar Wetland complex, Chennai, Tamil Nadu, southern India. Journal of Threatened Taxa 6(4): 5613-5635; http://dx.doi.org/10.11609/JoTT.o2905.5613-35

Shaji, C.P., Easa, P.S. & Gopalakrishnan, A. 2000. Freshwater Fish Diversity of Western Ghats. Pp. 33-55. In: Ponniah, A.G. and Gopalakrishnan, A. (eds.). Endemic Fish Diversity of Western Ghats. NBFGR – NATP Publication – 1,347 p. National Bureau of Fish Genetic Resources, Lucknow, U.P., India.

Sreenivasan, A. 1998. Fifty years of reservoir fisheries in Mettur dam, India: Some lessons. Naga The International Center for Living Aquatic Resources Management Quarterly 4-7.

Sreenivasan, A. 1976. Fish Production and Fish Population Changes in Some South Indian Reservoirs. Indian Journal of Fisheries 23(1 & 2), 134-152.

Talwar, P.K. & Jhingran, A.G. 1991. Inland Fishes of India and Adjacent Countries. Vol 1 & 2, Oxford and IBH Publishing Co., New Delhi, 1–1158 pp.

Table 1. Freshwater Fishes of Tamil Nadu with note on environment, endemic or exotic status, human use and conservation status

Order, Family, Scientific name	Authority	Environment			Endemic / Exotic	IUCN Status	Human Use	References
		F	B	M				
Order: Anguilliformes								
Family: Anguillidae								
Anguilla bengalensis	(Gray, 1831)	+	+	+		NT	Food	Knight & Devi (2010)
Anguilla bicolor	McClelland, 1844	+	+	+		NT	Food / Ornamental	Devi & Indra (2000); Knight & Devi (2010); Ramanujam *et al.* (2014)
Order: Beloniformes								
Family: Adrianichthyidae								
Oryzias carnaticus	(Jerdon, 1849)	+	+	-		LC	Ornamental	Shaji *et al.* (2000); Ramanujam *et al.* (2014); Mogalekar & Jawahar (2015)
Oryzias dancena	(Hamilton, 1822)	+	+	-		LC	Ornamental	Shaji *et al.* (2000); Knight & Devi (2010); Ramanujam *et al.* (2014); Mogalekar & Jawahar (2015)
Oryzias melastigma	(McClelland, 1839)	+	+	-		LC	Ornamental	Devi *et al.* (1999); Devi & Indra (2000); Mogalekar & Jawahar (2015)
Family: Belonidae								
Xenentodon cancila	(Hamilton, 1822)	+	+	+		LC	Ornamental / Food	Knight & Devi (2010); Ramanujam *et al.* (2014); Mogalekar & Jawahar (2015)
Family: Hemiramphidae								
Hyporhamphus limbatus	(Valenciennes, 1847)	+	+	+		LC	Food	Ramanujam *et al.* (2014); Mogalekar & Jawahar (2015)
Order: Clupeiformes								

Order, Family, Scientific name	Authority	Environment			Endemic / Exotic	IUCN Status	Human Use	References
		F	B	M				
Family: Clupeidae								
Hilsa kelee	(Cuvier, 1829)	+	+	+		NE	Food	Sreenivasan (1976); Talwar & Jhingran (1991)
Tenualosa ilisha	(Hamilton, 1822)	+	+	+		LC	Food	Sreenivasan (1998)
Order: Cypriniformes								
Family: Balitoridae								
Bhavania australis	(Jerdon, 1849)	+	-	-	ENWG	LC	Ornamental	Devi & Indra (2000); Johnson & Arunachalam (2009); Mogalekar & Jawahar (2015)
Family: Cobitidae								
Lepidocephalichthys guntea	(Hamilton, 1822)	+	-	-		LC	Ornamental	Knight & Devi (2010); Ramanujam *et al.* (2014); Mogalekar & Jawahar (2015)
Lepidocephalichthys thermalis	(Valenciennes, 1846)	+	-	-		LC	Ornamental	Indra (1992); Devi *et al.* (1999); Devi & Indra (2000); Balasundaram *et al.* (2001); Knight & Devi (2010); Ramanujam *et al.* (2014); Mogalekar & Jawahar (2015)
Family: Cyprinidae								
Amblypharyngodon melettinus	(Valenciennes, 1844)	+	-	-		LC	Ornamental	Devi & Indra (2000); Mogalekar & Jawahar (2015)
Amblypharyngodon microlepis	(Bleeker, 1853)	+	-	-		LC	Ornamental	Indra (1992); Devi *et al.* (1999); Devi & Indra (2000); Knight & Devi (2010); Ramanujam *et al.* (2014); Mogalekar & Jawahar (2015)
Amblypharyngodon mola	(Hamilton, 1822)	+	-	-		LC	Ornamental	Knight & Devi (2010); Mogalekar & Jawahar (2015)
Barbodes bovanicus	(Day, 1877)	+	-	-	ENCR	CR	Food	Devi & Indra (2000)
Barbodes carnaticus	(Jerdon, 1849)	+	-	-	ENI	LC	Food	Devi & Indra (2000)

Order, Family, Scientific name	Authority	Environment			Endemic / Exotic	IUCN Status	Human Use	References
		F	B	M				
Neolissochilus wynaadensis	(Day, 1873)	+	-	-	ENWG	CR	Ornamental	Devi & Indra (2000); Mogalekar & Jawahar (2015)
Barilius bakeri	Day, 1865	+	-	-	ENWG	LC	Ornamental	Jeyaraj (2000); Mogalekar & Jawahar (2015)
Barilius bendelisis	(Hamilton, 1807)	+	-	-		LC	Food	Devi & Indra (2000); Balasundaram *et al.* (2001); Mogalekar & Jawahar (2015)
Barilius gatensis	(Valenciennes, 1844)	+	-	-	ENWG	LC	Ornamental	Devi & Indra (2000); Mogalekar & Jawahar (2015)
Chela cachius	(Hamilton, 1822)	+	+	-		LC	Ornamental	Devi *et al.* (1999); Knight & Devi (2010); Mogalekar & Jawahar (2015)
Chela macrolepis	Knight & Devi, 2014	+	-	-	ENTN	NE	Ornamental	Knight & Devi (2014); Mogalekar & Jawahar (2015)
Laubuka dadiburjori	Menon, 1952	+	-	-	ENWG	LC	Ornamental	Shaji *et al.* (2000); Mogalekar & Jawahar (2015)
Laubuka fasciata	(Silas, 1958)	+	-	-	ENI	VU	Ornamental	Shaji *et al.* (2000); Mogalekar & Jawahar (2015)
Laubuka laubuca	(Hamilton, 1822)	+	+	-		LC	Ornamental	Devi *et al.* (1999); Knight & Devi (2010); Ramanujam *et al.* (2014); Mogalekar & Jawahar (2015)
Cirrhinus cirrhosus	(Bloch, 1795)	+	+	-	ENI	VU	Food	Knight & Devi (2010)
Cirrhinus fulungee	(Sykes, 1839)	+	-	-	ENWG	LC	Food	Shaji *et al.* (2000)
Cirrhinus macrops	Steindachner, 1870	+	-	-	ENI	NE	Food	Jayaram (2010)
Cirrhinus mrigala	(Hamilton, 1822)	+	-	-		LC	Food	Devi *et al.* (1999)

Order, Family, Scientific name	Authority	Environment			Endemic / Exotic	IUCN Status	Human Use	References
		F	B	M				
Cirrhinus reba	(Hamilton, 1822)	+	-	-		LC	Food	Balasundaram *et al.* (2001); Knight & Devi (2010)
Ctenopharyngodon idella	(Valenciennes, 1844)	+	-	-	Exotic	NE	Food	Knight & Devi (2010)
Cyprinus carpio var. communis	Linnaeus, 1758	+	-	-	Exotic	VU	Food	Devi & Indra (2000)
Cyprinus carpio var. specularis	Lacepède, 1803	+	-	-	Exotic	NE	Food	Devi & Indra (2000)
Danio rerio	(Hamilton, 1822)	+	-	-		LC	Ornamental	Devi *et al.* (1999); Devi & Indra (2000); Mogalekar & Jawahar (2015)
Dawkinsia arulius	(Jerdon, 1849)	+	-	-	ENCR	EN	Ornamental	Devi & Indra (2000); Mogalekar & Jawahar (2015)
Dawkinsia tambraparniei	(Silas, 1954)	+	-	-	ENTN	EN	Ornamental	Devi & Indra (2000); Johnson & Arunachalam (2009); Mogalekar & Jawahar (2015)
Dawkinsia filamentosa	(Valenciennes, 1844)	+	+	-	ENI	LC	Ornamental	Indra (1992); Devi *et al.* (1999); Johnson & Arunachalam (2009); Knight & Devi (2010); Ramanujam *et al.* (2014); Mogalekar & Jawahar (2015)
Devario aequipinnatus	(McClelland, 1839)	+	-	-		LC	Ornamental	Indra (1992); Johnson & Arunachalam (2009); Mogalekar & Jawahar (2015)
Devario neilgherriensis	(Day, 1867)	+	-	-	ENI	EN	Ornamental	Devi & Indra (2000); Mogalekar & Jawahar (2015)
Esomus barbatus	(Jerdon, 1849)	+	-	-	ENI	LC	Ornamental	Indra (1992); Knight & Devi (2010); Mogalekar & Jawahar (2015)
Esomus danricus	(Hamilton, 1822)	+	+	-		LC	Ornamental	Devi *et al.* (1999); Knight & Devi (2010); Ramanujam *et al.* (2014); Mogalekar & Jawahar (2015)

Order, Family, Scientific name	Authority	Environment			Endemic / Exotic	IUCN Status	Human Use	References
		F	B	M				
Esomus thermoicos	(Valenciennes, 1842)	+	-	-		LC	Ornamental	Devi *et al.* (1999); Knight & Devi (2010); Ramanujam *et al.* (2014); Mogalekar & Jawahar (2015)
Garra gotyla gotyla	(Gray, 1830)	+	-	-		LC	Ornamental / Food	Devi & Indra (2000); Dhinakaran *et al.* (2011); Mogalekar & Jawahar (2015)
Garra hughi	Silas, 1955	+	-	-	ENWG	EN	Ornamental	Devi & Indra (2000); Mogalekar & Jawahar (2015)
Garra kalakadensis	Rema Devi, 1993	+	-	-	ENTN	EN	Ornamental	Devi & Indra (2000); Dhinakaran *et al.* (2011); Mogalekar & Jawahar (2015)
Garra mcclellandi	(Jerdon, 1849)	+	-	-	ENCR	LC	Ornamental	Devi & Indra (2000); Mogalekar & Jawahar (2015)
Garra mullya	(Sykes, 1839)	+	-	-	ENI	LC	Ornamental	Indra (1992); Johnson & Arunachalam (2009); Dhinakaran *et al.* (2011); Mogalekar & Jawahar (2015)
Catla catla	(Hamilton, 1822)	+	+	-		LC	Food	Devi *et al.* (1999); Knight & Devi (2010); Ramanujam *et al.* (2014)
Haludaria afasciata	(Jayaram, 1990)	+	-	-		NE	Ornamental	Jayaram (2010); Mogalekar & Jawahar (2015)
Haludaria fasciata	(Jerdon, 1849)	+	-	-	ENSI	NE	Ornamental	Jayaram (2010); Mogalekar & Jawahar (2015)
Haludaria kannikattiensis	(Arunachalam & Johnson, 2003)	+	-	-	ENI	LC	Ornamental	Johnson & Arunachalam (2009); Mogalekar & Jawahar (2015)
Haludaria melanampyx	(Day, 1865)	+	-	-	ENWG	NE	Ornamental	Indra (1992); Devi & Indra (2000); Johnson & Arunachalam (2009); Mogalekar & Jawahar (2015)
Horadandia atukorali	Deraniyagala, 1943	+	+	-		LC	Ornamental	Devi *et al.* (1999); Devi & Indra (2000); Knight & Devi (2010); Mogalekar & Jawahar (2015)

Order, Family, Scientific name	Authority	Environment			Endemic / Exotic	IUCN Status	Human Use	References
		F	B	M				
Horalabiosa joshuai	Silas, 1954	+	-	-	ENTN	EN	Food / Ornamental	Devi & Indra (2000); Johnson & Arunachalam (2009); Mogalekar & Jawahar (2015)
Horalabiosa palaniensis	Devi & Menon, 1994	+	-	-	ENTN	VU	Ornamental	Devi & Indra (2000); Mogalekar & Jawahar (2015)
Hypophthalmichthys molitrix	(Valenciennes, 1844)	+	-	-	Exotic	NT	Food	Devi & Indra (2000); Knight & Devi (2010)
Hypselobarbus dubius	(Day, 1867)	+	-	-	ENTN	EN	Food	Johnson & Arunachalam (2009)
Hypselobarbus curmuca	(Hamilton, 1807)	+	-	-	ENI	EN	Food	Johnson & Arunachalam (2009)
Hypselobarbus dobsoni	(Day, 1876)	+	-	-	ENWG	DD	Food	Devi & Indra (2000); Johnson & Arunachalam (2009)
Hypselobarbus jerdoni	(Day, 1870)	+	-	-	ENWG	LC	Food	Devi & Indra (2000)
Hypselobarbus kurali	Menon & Devi, 1995	+	-	-	ENTN	LC	Food	Devi & Indra (2000)
Hypselobarbus lithopidos	(Day, 1874)	+	-	-	ENI	DD	Food	Jayaram (2010)
Hypselobarbus micropogon	(Valenciennes, 1842)	+	-	-	ENWG	EN	Food	Shaji *et al.* (2000)
Bangana ariza	(Hamilton, 1807)	+	-	-		LC	Food	Devi & Indra (2000)
Bangana dero	(Hamilton, 1822)	+	-	-		LC	Food	Devi & Indra (2000); Ramanujam *et al.* (2014)
Labeo altivelis	Peters, 1852	+	-	-	Exotic	LC	Food	Sreenivasan (1998)
Labeo bata	(Hamilton, 1822)	+	-	-		LC	Food	Devi & Indra (2000)
Labeo boga	(Hamilton, 1822)	+	-	-		LC	Food	Devi & Indra (2000)
Labeo boggut	(Sykes, 1839)	+	-	-		LC	Food	Devi & Indra (2000)

Order, Family, Scientific name	Authority	Environment			Endemic / Exotic	IUCN Status	Human Use	References
		F	B	M				
Labeo calbasu	(Hamilton, 1822)	+	+	-		LC	Food	Balasundaram *et al.* (2001); Knight & Devi (2010)
Labeo fimbriatus	(Bloch, 1795)	+	-	-		LC	Food	Sreenivasan (1998)
Labeo kontius	(Jerdon, 1849)	+	-	-	ENCR	LC	Food	Sreenivasan (1998)
Labeo pangusia	(Hamilton, 1822)	+	-	-		NT	Food	Devi & Indra (2000)
Labeo potail	(Sykes, 1839)	+	-	-	ENCR	EN	Food	Devi & Indra (2000); Mogalekar & Jawahar (2015)
Labeo rohita	(Hamilton, 1822)	+	+	-		LC	Food	Devi *et al.* (1999); Knight & Devi (2010)
Osteobrama cotio	(Hamilton, 1822)	+	-	-		LC	Ornamental	Devi & Indra (2000); Mogalekar & Jawahar (2015)
Osteobrama neilli	(Day, 1873)	+	-	-	ENTN	LC	Ornamental	Devi & Indra (2000); Mogalekar & Jawahar (2015)
Osteobrama peninsularis	Silas 1952	+	-	-	ENI	DD	Food / Ornamental	Knight & Devi (2010); Mogalekar & Jawahar (2015)
Osteochilichthys brevidorsalis	(Day, 1873)	+	-	-	ENI	LC	Food	Devi & Indra (2000)
Osteochilus nashii	(Day, 1869)	+	-	-	ENI	LC	Ornamental	Devi & Indra (2000); Mogalekar & Jawahar (2015)
Pethia conchonius	(Hamilton, 1822)	+	-	-		LC	Ornamental	Devi *et al.* (1999); Devi & Indra (2000); Balasundaram *et al.* (2001); Knight & Devi (2010); Ramanujam *et al.* (2014); Mogalekar & Jawahar (2015)
Pethia gelius	(Hamilton, 1822)	+	-	-		LC	Ornamental	Knight (2010); Mogalekar & Jawahar (2015)
Pethia guganio	(Hamilton, 1822)	+	-	-		LC	Ornamental	Shaji *et al.* (2000); Mogalekar & Jawahar (2015)

Order, Family, Scientific name	Authority	Environment			Endemic / Exotic	IUCN Status	Human Use	References
		F	B	M				
Pethia nigripinna	(Knight, Devi, Indra & Arunachalam, 2012)	+	-	-	ENWG	NE	Ornamental	Knight *et al.* (2012); Mogalekar & Jawahar (2015)
Pethia punctata	(Day, 1865)	+	-	-		LC	Ornamental	Devi & Indra (2000); Mogalekar & Jawahar (2015)
Pethia sharmai	(Menon & Devi, 1993)	+	-	-	ENTN	EN	Ornamental	Devi *et al.* (1999); Devi & Indra (2000); Knight & Devi (2010); Ramanujam *et al.* (2014); Mogalekar & Jawahar (2015)
Pethia ticto	(Hamilton, 1822)	+	+	-		LC	Ornamental	Devi *et al.* (1999); Johnson & Arunachalam (2009); Knight & Devi (2010); Ramanujam *et al.* (2014); Mogalekar & Jawahar (2015)
Puntius amphibius	(Valenciennes, 1842)	+	+	-		DD	Ornamental	Indra (1992); Devi *et al.* (1999); Devi & Indra (2000); Johnson & Arunachalam (2009); Knight & Devi (2010); Ramanujam *et al.* (2014); Mogalekar & Jawahar (2015)
Puntius arenatus	(Day, 1878)	+	-	-	ENI	VU	Ornamental	Devi & Indra (2000); Mogalekar & Jawahar (2015)
Puntius bimaculatus	(Bleeker, 1863)	+	-	-		LC	Ornamental	Devi & Indra (2000); Johnson & Arunachalam (2009); Mogalekar & Jawahar (2015)
Puntius cauveriensis	(Hora, 1937)	+	-	-	ENCR	EN	Ornamental	Jayaram (2010); Mogalekar & Jawahar (2015)
Puntius chola	(Hamilton, 1822)	+	-	-		LC	Ornamental	Indra (1992); Devi *et al.* (1999); Devi & Indra (2000); Knight & Devi (2010); Ramanujam *et al.* (2014); Mogalekar & Jawahar (2015)

Order, Family, Scientific name	Authority	Environment			Endemic / Exotic	IUCN Status	Human Use	References
		F	B	M				
Puntius dorsalis	(Jerdon, 1849)	+	-	-		LC	Ornamental / Food	Indra (1992); Devi *et al.* (1999); Devi & Indra (2000); Johnson & Arunachalam (2009); Knight & Devi (2010); Ramanujam *et al.* (2014); Mogalekar & Jawahar (2015)
Puntius mahecola	(Valenciennes, 1844)	+	-	-	ENKL	DD	Ornamental	Knight & Devi (2010); Ramanujam *et al.* (2014); Mogalekar & Jawahar (2015)
Puntius melanostigma	(Day, 1878)	+	-	-	ENI	NE	Ornamental	Devi & Indra (2000); Mogalekar & Jawahar (2015)
Puntius mudumalaiensis	Menon & Devi, 1992	+	-	-	ENTN	VU	Ornamental	Devi & Indra (2000); Mogalekar & Jawahar (2015)
Puntius parrah	Day, 1865	+	-	-	ENWG	LC	Ornamental	Devi & Indra (2000); Mogalekar & Jawahar (2015)
Puntius sophore	(Hamilton, 1822)	+	+	-		LC	Ornamental	Indra (1992); Devi *et al.* (1999); Balasundaram *et al.* (2001); Knight & Devi (2010); Ramanujam *et al.* (2014); Mogalekar & Jawahar (2015)
Puntius thomassi	(Day, 1874)	+	-	-	ENWG	CR	Food / Ornamental	Talwar & Jhingran (1991)
Puntius vittatus	Day, 1865	+	+	-		LC	Ornamental	Indra (1992); Devi *et al.* (1999); Devi & Indra (2000); Knight & Devi (2010); Ramanujam *et al.* (2014); Mogalekar & Jawahar (2015)
Rasbora caverii	(Jerdon, 1849)	+	+	-		LC	Ornamental	Devi *et al.* (1999); Knight & Devi (2010); Ramanujam *et al.* (2014); Mogalekar & Jawahar (2015)
Rasbora daniconius	(Hamilton, 1822)	+	+	-		LC	Ornamental	Indra (1992); Devi *et al.* (1999); Balasundaram *et al.* (2001); Johnson & Arunachalam (2009); Knight & Devi (2010); Ramanujam *et al.* (2014); Mogalekar & Jawahar (2015)

Order, Family, Scientific name	Authority	Environment			Endemic / Exotic	IUCN Status	Human Use	References
		F	B	M				
Rasbora rasbora	(Hamilton, 1822)	+	+	-		LC	Ornamental	Jayaram (2010); Mogalekar & Jawahar (2015)
Salmostoma acinaces	(Valenciennes, 1844)	+	-	-		LC	Ornamental / Food	Knight & Devi (2010); Mogalekar & Jawahar (2015)
Salmostoma bacaila	(Hamilton, 1822)	+	+	-		LC	Ornamental / Food	Indra (1992); Knight & Devi (2010); Mogalekar & Jawahar (2015)
Salmostoma balookee	(Sykes, 1839)	+	-	-		LC	Food	Indra (1992); Devi *et al.* (1999); Devi & Indra (2000); Knight & Devi (2010); Ramanujam *et al.* (2014)
Salmostoma novacula	(Valenciennes, 1840)	+	-	-	ENI	LC	Ornamental	Talwar & Jhingran (1991); Devi & Indra (2000); Mogalekar & Jawahar (2015)
Salmostoma orissaensis	(Banarescu, 1968)	+	-	-	ENI	NE	Ornamental / Food	Talwar & Jhingran (1991); Mogalekar & Jawahar (2015)
Salmostoma untrahi	(Day, 1869)	+	-	-	ENI	LC	Food	Talwar & Jhingran (1991)
Systomus rubripinnis	(Valenciennes 1842)	+	-	-	Exotic	DD	Food	Knight & Devi (2010)
Systomus sarana	(Hamilton, 1822)	+	+	-		LC	Food / Ornamental	Indra (1992); Devi *et al.* (1999); Devi & Indra (2000); Balasundaram *et al.* (2001); Knight & Devi (2010); Ramanujam *et al.* (2014); Mogalekar & Jawahar (2015)
Tinca tinca	(Linnaeus, 1758)	+	+	-	Exotic	LC	Food / Ornamental	Sreenivasan (1998); Mogalekar & Jawahar (2015)
Tor khudree	(Sykes, 1839)	+	-	-		EN	Food	Johnson & Arunachalam (2009)
Tor malabaricus	(Jerdon, 1849)	+	-	-	ENI	EN	Food	Indra (1992)
Tor putitora	(Hamilton, 1822)	+	-	-		EN	Food	Sreenivasan (1998)
Tor tor	(Hamilton, 1822)	+	-	-		NT	Food	Sreenivasan (1998)

Order, Family, Scientific name	Authority	Environment			Endemic / Exotic	IUCN Status	Human Use	References
		F	B	M				
Family: Nemacheilidae								
Paracanthocobitis mooreh	(Sykes, 1839)	+	-	-	ENPI	LC	Ornamental	Devi & Indra (2000); Mogalekar & Jawahar (2015)
Nemacheilus denisoni	Day, 1867	+	-	-	ENI	LC	Ornamental	Devi & Indra (2000); Mogalekar & Jawahar (2015)
Nemacheilus kodaguensis	Menon, 1987	+	-	-	ENCR	VU	Ornamental	Jayaram (2010); Mogalekar & Jawahar (2015)
Nemacheilus nilgiriensis	Menon, 1987	+	-	-	ENTN	LC	Ornamental	Gopalakrishnan & Ponniah (2000); Mogalekar & Jawahar (2015)
Nemacheilus pulchellus	Day, 1873	+	-	-	ENWG	EN	Ornamental	Gopalakrishnan & Ponniah (2000) ; Mogalekar & Jawahar (2015)
Nemacheilus semiarmatus	Day, 1867	+	-	-	ENPI	LC	Ornamental	Gopalakrishnan & Ponniah (2000); Mogalekar & Jawahar (2015)
Nemacheilus triangularis	Day, 1865	+	-	-	ENTN, ENKL	LC	Ornamental	Indra (1992); Johnson & Arunachalam (2009); Mogalekar & Jawahar (2015)
Order: Cyprinodontiformes								
Family: Aplocheilidae								
Aplocheilus blockii	Arnold, 1911	+	+	-		LC	Ornamental	Indra (1992); Devi *et al.* (1999); Gopalakrishnan & Ponniah (2000); Mogalekar & Jawahar (2015)
Aplocheilus lineatus	(Valenciennes, 1846)	+	+	-	ENPI	LC	Ornamental	Indra (1992); Mogalekar & Jawahar (2015)
Aplocheilus parvus	(Sundara Raj, 1916)	+	+	-		NE	Ornamental	Knight & Devi (2010); Ramanujam *et al.* (2014); Mogalekar & Jawahar (2015)
Family: Poeciliidae								

Order, Family, Scientific name	Authority	Environment			Endemic / Exotic	IUCN Status	Human Use	References
		F	B	M				
Gambusia affinis	(Baird & Girard, 1853)	+	+	-	Exotic	LC	Ornamental	Indra (1992); Devi *et al.* (1999); Devi & Indra (2000); Knight & Devi (2010); Ramanujam *et al.* (2014); Mogalekar & Jawahar (2015)
Poecilia reticulata	Peters, 1859	+	+	-	Exotic	NE	Ornamental	Devi *et al.* (1999); Devi & Indra (2000); Ramanujam *et al.* (2014); Mogalekar & Jawahar (2015)
Xiphophorus hellerii	Heckel, 1848	+	+	-	Exotic	NE	Ornamental	Devi *et al.* (1999)
Xiphophorus maculatus	(Günther, 1866)	+	-	-	Exotic	NE	Ornamental	Devi *et al.* (1999)
Order: Elopiformes								
Family: Megalopidae								
Megalops cyprinoides	(Broussonet, 1782)	+	+	+		DD	Food	Indra (1992); Devi & Indra (2000); Ramanujam *et al.* (2014)
Order: Gonorynchiformes								
Family: Chanidae								
Chanos chanos	(Forsskål, 1775)	+	+	+		NE	Food	Sreenivasan (1998)
Order: Mugiliformes								
Family: Mugilidae								
Chelon parsia	(Hamilton, 1822)	+	+	+		NE	Food	Indra (1992); Devi & Indra (2000)
Chelon planiceps	(Valenciennes, 1836)	+	+	+		DD	Food	Devi & Indra (2000)
Chelon subviridis	(Valenciennes, 1836)	+	+	+		NE	Food	Devi & Indra (2000)
Mugil cephalus	Linnaeus, 1758	+	+	+		LC	Food	Indra (1992); Devi & Indra (2000)

Order, Family, Scientific name	Authority	Environment			Endemic / Exotic	IUCN Status	Human Use	References
		F	B	M				
Rhinomugil corsula	(Hamilton, 1822)	+	+	-		LC	Food	Sreenivasan (1998)
Sicamugil cascasia	(Hamilton, 1822)	+	-	-		LC	Food	Gopalakrishnan & Ponniah (2000)
Order: Osteoglossiformes								
Family: Notopteridae								
Notopterus notopterus	(Pallas, 1769)	+	-	-		LC	Food / Ornamental	Devi *et al.* (1999); Knight & Devi (2010); Ramanujam *et al.* (2014); Mogalekar & Jawahar (2015)
Order: Perciformes								
Family: Ambassidae								
Ambassis ambassis	(Lacepède, 1802)	+	+	+		LC	Ornamental	Indra (1992); Devi & Indra (2000); Mogalekar & Jawahar (2015)
Ambassis gymnocephalus	(Lacepède, 1802)	+	+	+		LC	Ornamental	Jayaram (2010); Mogalekar & Jawahar (2015)
Ambassis interrupta	Bleeker, 1853	+	+	+		LC	Ornamental	Talwar & Jhingran (1991); Mogalekar & Jawahar (2015)
Ambassis miops	Günther, 1872	+	+	+		LC	Ornamental	Talwar & Jhingran (1991); Mogalekar & Jawahar (2015)
Chanda nama	Hamilton, 1822	+	+	-		LC	Ornamental	Devi *et al.* (1999); Knight & Devi (2010); Ramanujam *et al.* (2014); Mogalekar & Jawahar (2015)
Parambassis lala	(Hamilton 1822)	+	+	-		NT	Ornamental	Knight & Devi (2010); Mogalekar & Jawahar (2015)
Parambassis ranga	(Hamilton, 1822)	+	+	-		LC	Ornamental	Devi *et al.* (1999); Knight & Devi (2010); Ramanujam *et al.* (2014); Mogalekar & Jawahar (2015)

Order, Family, Scientific name	Authority	Environment			Endemic / Exotic	IUCN Status	Human Use	References
		F	B	M				
Family: Anabantidae								
Anabas testudineus	(Bloch, 1792)	+	+	-		DD	Ornamental / Food	Devi *et al.* (1999); Knight & Devi (2010); Ramanujam *et al.* (2014); Mogalekar & Jawahar (2015)
Family: Badidae								
Badis badis	(Hamilton 1822)	+	-	-		LC	Ornamental / Food	Knight & Devi (2009); Mogalekar & Jawahar (2015)
Family: Channidae								
Channa gachua	(Hamilton 1822)	+	-	-		LC	Ornamental	Knight & Devi (2010); Mogalekar & Jawahar (2015)
Channa punctata	(Bloch, 1793)	+	+	-		LC	Food	Indra (1992); Devi *et al.* (1999); Knight & Devi (2010); Ramanujam *et al.* (2014); Mogalekar & Jawahar (2015)
Channa striata	(Bloch, 1793)	+	+	-		LC	Food	Devi *et al.* (1999); Knight & Devi (2010); Ramanujam *et al.* (2014); Mogalekar & Jawahar (2015)
Family: Cichlidae								
Cichlasoma trimaculatum	(Gunther 1867)	+	-	-	Exotic	NE	Ornamental	Knight & Devi (2010); Mogalekar & Jawahar (2015)
Etroplus maculatus	(Bloch, 1795)	+	+	-		LC	Ornamental / Food	Indra (1992); Devi *et al.* (1999); Balasundaram *et al.* (2001); Knight & Devi (2010); Ramanujam *et al.* (2014); Mogalekar & Jawahar (2015)
Etroplus suratensis	(Bloch, 1790)	+	+	-		LC	Ornamental / Food	Indra (1992); Balasundaram *et al.* (2001); Knight & Devi (2010); Mogalekar & Jawahar (2015)

Order, Family, Scientific name	Authority	Environment			Endemic / Exotic	IUCN Status	Human Use	References
		F	B	M				
Hemichromis bimaculatus	Gill, 1862	+	+	-	Exotic	LC	Ornamental / Food	Knight & Devi (2010); Mogalekar & Jawahar (2015)
Oreochromis aureus	(Steindachner, 1864)	+	+	-	Exotic	LC	Food	Knight & Devi (2010); Mogalekar & Jawahar (2015)
Oreochromis mossambicus	(Peters, 1852)	+	+	-	Exotic	LC	Food	Indra (1992); Devi *et al.* (1999); Balasundaram *et al.* (2001); Knight & Devi (2010); Ramanujam *et al.* (2014); Mogalekar & Jawahar (2015)
Oreochromis niloticus	(Linnaeus, 1758)	+	+	-	Exotic	NE	Food	Knight & Devi (2010)
Family: Gobiidae								
Amblyeleotris gymnocephala	(Bleeker, 1853)	+	+	-		NE	Ornamental	Chatterjee *et al.* (2013); Mogalekar & Jawahar (2015)
Apocryptodon madurensis	(Bleeker, 1849)	+	+	+		NE	Ornamental	Talwar & Jhingran (1991); Mogalekar & Jawahar (2015)
Awaous grammepomus	(Bleeker, 1849)	+	+	-		LC	Ornamental / Food	Indra (1992); Devi & Indra (2000); Mogalekar & Jawahar (2015)
Favonigobius reichei	(Bleeker, 1854)	+	+	+		LR/nt	Food	Devi *et al.* (1999)
Glossogobius giuris	(Hamilton, 1822)	+	+	+		LC	Ornamental / Food	Indra (1992); Devi *et al.* (1999); Balasundaram *et al.* (2001); Knight & Devi (2010); Ramanujam *et al.* (2014); Mogalekar & Jawahar (2015)
Oligolepis acutipennis	(Valenciennes, 1837)	+	+	+		DD	Ornamental	Indra (1992)

Order, Family, Scientific name	Authority	Environment			Endemic / Exotic	IUCN Status	Human Use	References
		F	B	M				
Oxuderces dentatus	Eydoux & Souleyet, 1850	+	+	-		NE	Ornamental	Jayaram (2010); Mogalekar & Jawahar (2015)
Psammogobius biocellatus	(Valenciennes, 1837)	+	+	+		LC	Ornamental	Indra (1992)
Pseudogobius javanicus	(Bleeker, 1856)	+	+	+		NE	Ornamental	Indra (1992)
Sicyopterus griseus	(Day, 1877)	+	+	-	ENI	LC*	Ornamental	Arunachalam & Sankaranarayanan (1999); Jayaram (2010); Mogalekar & Jawahar (2015); Mogalekar & Jawahar (2015)
Family: Nandidae								
Nandus nandus	(Hamilton, 1822)	+	+	-		LC	Ornamental / Food	Knight & Devi (2010); Mogalekar & Jawahar (2015)
Family: Osphronemidae								
Osphronemus gourami	(Lacepède 1801)	+	+	-	Exotic	LC	Ornamental	Knight & Devi (2010); Ramanujam *et al.* (2014); Mogalekar & Jawahar (2015)
Pseudosphromenus cupanus	(Cuvier, 1831)	+	+	-		LC	Ornamental	Indra (1992); Devi *et al.* (1999); Knight & Devi (2010); Mogalekar & Jawahar (2015)
Trichogaster fasciata	Bloch & Schneider, 1801	+	-	-		LC	Ornamental	Devi *et al.* (1999)
Trichogaster lalius	(Hamilton, 1822)	+	-	-		LC	Ornamental	Knight & Devi (2010); Ramanujam *et al.* (2014); Mogalekar & Jawahar (2015)
Trichopodus trichopterus	(Pallas, 1770)	+	-	-	Exotic	LC	Ornamental	Knight & Devi (2010); Ramanujam *et al.* (2014); Mogalekar & Jawahar (2015)
Family: Pristolepididae								
Pristolepis marginata	Jerdon, 1849	+	-	-	ENWG	LC	Ornamental / Food	Jayaram (2010); Mogalekar & Jawahar (2015)

Order, Family, Scientific name	Authority	Environment			Endemic / Exotic	IUCN Status	Human Use	References
		F	B	M				
Order: Salmoniformes								
Family: Salmonidae								
Oncorhynchus mykiss	(Walbaum, 1792)	+	+	+	Exotic	NE	Food	Sreenivasan (1998); Jayaram (2010)
Oncorhynchus nerka	(Walbaum, 1792)	+	+	+	Exotic	LC	Food	Jayaram (2010)
Order: Siluriformes								
Family: Bagridae								
Hemibagrus punctatus	(Jerdon, 1849)	+	-	-	ENCR	CR	Ornamental / Food	Devi & Indra (2000); Mogalekar & Jawahar (2015)
Mystus armatus	(Day, 1865)	+	+	-		LC	Ornamental / Food	Indra (1992); Devi & Indra (2000); Rajagopal & Davidar (2013); Mogalekar & Jawahar (2015)
Mystus bleekeri	(Day, 1877)	+	-	-		LC	Ornamental / Food	Devi *et al.* (1999); Devi & Indra (2000); Balasundaram *et al.* (2001); Knight & Devi (2010); Ramanujam *et al.* (2014); Mogalekar & Jawahar (2015)
Mystus cavasius	(Hamilton, 1822)	+	+	-		LC	Food	Devi *et al.* (1999); Devi & Indra (2000); Balasundaram *et al.* (2001); Knight & Devi (2010); Ramanujam *et al.* (2014)
Mystus gulio	(Hamilton, 1822)	+	+	-		LC	Food	Devi *et al.* (1999); Devi & Indra (2000); Knight & Devi (2010); Rajagopal & Davidar (2013); Ramanujam *et al.* (2014)
Mystus keletius	(Valenciennes, 1840)	+	-	-		LC	Ornamental	Devi & Indra (2000); Knight & Devi (2010); Rajagopal & Davidar (2013); Mogalekar & Jawahar (2015)

Order, Family, Scientific name	Authority	Environment			Endemic / Exotic	IUCN Status	Human Use	References
		F	B	M				
Mystus malabaricus	(Jerdon, 1849)	+	+	-	ENWG	NT	Ornamental	Devi & Indra (2000); Mogalekar & Jawahar (2015)
Mystus montanus	(Jerdon, 1849)	+	+	-	ENI	LC	Ornamental	Indra (1992); Devi & Indra (2000); Rajagopal & Davidar (2013); Mogalekar & Jawahar (2015)
Mystus oculatus	(Valenciennes, 1840)	+	+	-	ENTN, ENKL	LC	Food	Indra (1992); Devi & Indra (2000)
Mystus seengtee	(Sykes, 1839)	+	-	-	ENSI	LC	Food	Rajagopal & Davidar (2013)
Mystus vittatus	(Bloch, 1794)	+	+	-		LC	Ornamental / Food	Indra (1992); Devi *et al.* (1999); Devi & Indra (2000); Balasundaram *et al.* (2001); Knight & Devi (2010); Rajagopal & Davidar (2013); Ramanujam *et al.* (2014); Mogalekar & Jawahar (2015)
Sperata aor	(Hamilton, 1822)	+	-	-		LC	Food	Devi & Indra (2000); Knight & Devi (2010)
Sperata seenghala	(Sykes, 1839)	+	+	-		LC	Food	Devi & Indra (2000)
Family: Clariidae								
Clarias batrachus	(Linnaeus, 1758)	+	+	-		LC	Food	Balasundaram *et al.* (2001)
Clarias dayi	Hora, 1936	+	-	-	ENTN, ENKL	NE	Food	Arunachalam & Manimekalan (2000)
Clarias dussumieri	Valenciennes, 1840	+	-	-	ENSI	NT	Food	Devi & Indra (2000)
Clarias magur	(Hamilton 1822)	+	-	-		EN	Food	Knight & Devi (2010)
Clarias gariepinus	(Burchell 1822)	+	-	-	Exotic	LC	Food	Knight & Devi (2010)
Family: Heteropneustidae								

Order, Family, Scientific name	Authority	Environment			Endemic / Exotic	IUCN Status	Human Use	References
		F	B	M				
Heteropneustes fossilis	(Bloch, 1794)	+	+	-		LC	Food	Indra (1992); Devi *et al.* (1999); Knight & Devi (2010); Rajagopal & Davidar (2013); Ramanujam *et al.* (2014)
Heteropneustes longipectoralis	Devi & Raghunathan, 1999	+	-	-	ENTN	DD	Ornamental / Food	Devi & Raghunathan (1999); Jayaram (2010); Mogalekar & Jawahar (2015)
Heteropneustes microps	(Günther, 1864)	+	+	-		NE	Ornamental / Food	Devi & Indra (2000); Mogalekar & Jawahar (2015)
Family: Loricariidae								
Pterygoplichthys disjunctivus	(Weber, 1991)	+	-	-	Exotic	NE	Ornamental / Food	Knight & Devi (2010); Mogalekar & Jawahar (2015)
Pterygoplichthys pardalis	(Castelnau, 1855)	+	-	-	Exotic	NE	Ornamental / Food	Knight & Devi (2010); Mogalekar & Jawahar (2015)
Family: Pangasiidae								
Pangasius pangasius	(Hamilton, 1822)	+	+	-		LC	Food	Knight & Devi (2010)
Family: Schilbeidae								
Neotropius atherinoides	(Bloch, 1794)	+	+	-		LC	Ornamental / Food	Devi *et al.* (1999); Balasundaram *et al.* (2001); Knight & Devi (2010); Rajagopal & Davidar (2013); Ramanujam *et al.* (2014); Mogalekar & Jawahar (2015)
Silonia silondia	(Hamilton, 1822)	+	+	-		LC	Food	Molur & Walker (1998)
Family: Siluridae								
Ompok bimaculatus	(Bloch, 1794)	+	+	-		NT	Food	Balasundaram *et al.* (2001); Knight & Devi (2010); Rajagopal & Davidar (2013); Mogalekar & Jawahar (2015)
Ompok karunkodu	Ng, 2013	+	-	-	ENCR	NE	Food	Ng (2013)

Order, Family, Scientific name	Authority	Environment			Endemic / Exotic	IUCN Status	Human Use	References
		F	B	M				
Ompok malabaricus	(Valenciennes, 1840)	+	-	-	ENWG	LC	Food	Rajagopal & Davidar (2013)
Pterocryptis wynaadensis	(Day, 1873)	+	-	-	ENWG	EN	Food	Shaji *et al.* (2000); Johnson & Arunachalam (2009)
Wallago attu	(Bloch & Schneider, 1801)	+	+	-		NT	Food	Knight & Devi (2010); Ramanujam *et al.* (2014)
Family: Sisoridae								
Glyptothorax annandalei	Hora, 1923	+	-	-		LC	Ornamental	Jayaram (2010); Mogalekar & Jawahar (2015)
Glyptothorax housei	Herre, 1942	+	-	-	ENWG	EN	Ornamental	Jayaram (2010); Mogalekar & Jawahar (2015)
Glyptothorax madraspatanum	(Day, 1873)	+	-	-	ENCR	EN	Ornamental	Shaji *et al.* (2000); Johnson & Arunachalam (2009); Mogalekar & Jawahar (2015)
Order: Synbranchiformes								
Family: Mastacembelidae								
Macrognathus aral	(Bloch & Schneider, 1801)	+	+	-		LC	Ornamental / Food	Devi *et al.* (1999); Knight & Devi (2010); Ramanujam *et al.* (2014); Mogalekar & Jawahar (2015)
Macrognathus guentheri	(Day, 1865)	+	-	-	ENI	LC	Ornamental / Food	Indra (1992); Devi & Indra (2000); Mogalekar & Jawahar (2015)
Macrognathus pancalus	Hamilton, 1822	+	+	-		LC	Ornamental / Food	Devi *et al.* (1999); Knight & Devi (2010); Ramanujam *et al.* (2014); Mogalekar & Jawahar (2015)

Order, Family, Scientific name	Authority	Environment			Endemic / Exotic	IUCN Status	Human Use	References
		F	B	M				
Mastacembelus armatus	(Lacepède, 1800)	+	+	-		LC	Ornamental / Food	Devi *et al.* (1999); Balasundaram *et al.* (2001); Johnson & Arunachalam (2009); Knight & Devi (2010); Mogalekar & Jawahar (2015)
Order: Syngnathiformes								
Family: Syngnathidae								
Hippichthys penicillus	(Cantor, 1849)	+	+	+		LC	Ornamental	Shaji *et al.* (2000); Mogalekar & Jawahar (2015)
Microphis cuncalus	(Hamilton, 1822)	+	+	-		LC	Ornamental	Talwar & Jhingran (1991); Jayaram (2010); Mogalekar & Jawahar (2015)

F = freshwater; B = brackish; M = Marine; + = Present; - = Absent; ENI = endemic to India; ENPI = endemic to peninsular India; ENSI = endemic to south India; ENWG = endemic to Western Ghats of India; ENCR = endemic to Cauvery River; ENTN = endemic to Tamil Nadu; ENKL = endemic to Kerala; IUCN = International Union for Conservation of Nature; CR = Critically Endangered; EN= Endangered; NT = Near Threatened; VU = Vulnerable; LR/nt = Lower Risk- Near Threatened LC = Least Concern; DD = Data Deficient; NE = Not Evaluated

Table 2. Number of family, genera and species under each Order of Freshwater Fishes of Tamil Nadu

Order	Family	Number of Genera	Number of Species
Anguilliformes	Anguillidae	1	2
Beloniformes	Adrianichthyidae	1	3
	Belonidae	1	1
	Hemiramphidae	1	1
Clupeiformes	Clupeidae	1	2
Cypriniformes	Balitoridae	1	1
	Cobitidae	1	2
	Cyprinidae	32	105
	Nemacheilidae	2	7
Cyprinodontiformes	Aplocheilidae	1	3
	Poeciliidae	3	4
Elopiformes	Megalopidae	1	1
Gonorynchiformes	Chanidae	1	1
Mugiliformes	Mugilidae	5	6
Osteoglossiformes	Notopteridae	1	1
Perciformes	Ambassidae	3	7
	Anabantidae	1	1
	Badidae	1	1
	Channidae	1	3
	Cichlidae	4	7
	Gobiidae	9	10
	Nandidae	1	1
	Osphronemidae	3	5
	Pristolepididae	1	1
Salmoniformes	Salmonidae	1	2
Siluriformes	Bagridae	3	13
	Clariidae	1	5
	Heteropneustidae	1	3
	Loricariidae	1	2
	Pangasiidae	1	1
	Schilbeidae	2	2
	Siluridae	3	5
	Sisoridae	1	3
Synbranchiformes	Mastacembelidae	2	4
Syngnathiformes	Syngnathidae	2	2
Total = 14	**Total = 35**	**Total = 95**	Total = 218

Figure 1. Number of genera and species under each family

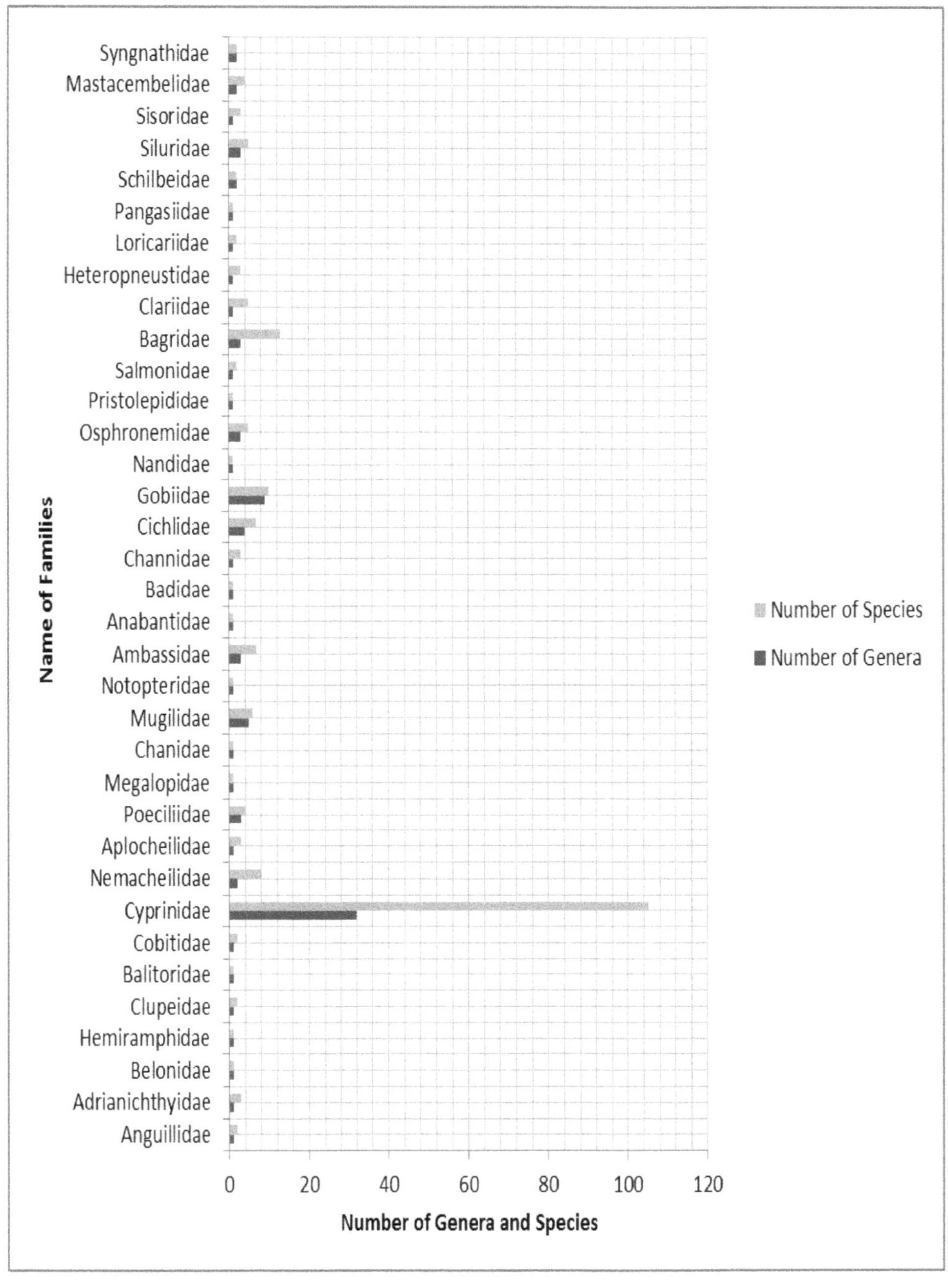

5

FRESHWATER FISH DIVERSITY OF UNITED ANDHRA PRADESH

***H.S. Mogalekar*[1]*, Prateek*[1]*, N.A. Ingole*[2] *and D.S. Patadiya*[1]**

[1] *Department of Fisheries Biology and Resource Management, Fisheries College and Research Institute, Tamil Nadu Fisheries University, Thoothukudi-628 008, Tamil Nadu, India*

[2] *Department of Fisheries Resource Management, College of Fisheries, G.B. Pant University of Agriculture and Technology, Pantnagar-263 145, Uttarakhand, India*

INTRODUCTION

United Andhra Pradesh is enriched with a vast expanse of freshwater fisheries resources. The Eastern Ghats to the east coastal line, the Deccan plateau with the Sahyadri range to the north border and the Horseley and other hills to the south border are three major mountain ranges located in the state. Godavari, Krishna and Pennar Rivers with their several tributaries form the chief perennial river systems of the state (Barman, 2009). Moreover, United Andhra Pradesh has 98 small reservoirs, 2 800 tanks, 32 medium reservoir and 7 large reservoirs with a total surface water area of 458 507 ha. The seven large reservoirs have a total surface area of 190 151 ha, while the medium and small reservoirs cover 66 429 ha and 24 178 ha respectively (Sugunan, 1995).

A review of literature on the fish fauna of Andhra Pradesh shows that several works have been carried out on the freshwater fish fauna of United Andhra Pradesh by Sykes (1838); Jerdon (1849); Misra (1938); Rahimullah (1943 & 1944); Mahmood and Rahimullah (1947); Chacko (1949); David (1963); Murthy (1977); Rao and Reddy (1984); Barman (1993); Devi and Indra (2003); Chandrasekhar (2004); Devi *et al.* (2005); Barman (2009); Srikanth *et al.* (2009); Aruna *et al.* (2011); Rao *et al.* (2011);

Ramulu and Benarjee (2013); Rao *et al.* (2013); Reddy *et al.* (2013); Thirupathaiah *et al.* (2013); Rao (2014a); Rao (2014b); Thirupathaiah *et al.* (2014); Rao (2015) and Rao *et al.* (2015). Updated checklist of freshwater fish fauna of United Andhra Pradesh is not available now. Henceforth present checklist is intended to update the scientific nomenclature of all the fish species recorded from freshwater bodies of United Andhra Pradesh to assist fish taxonomists in future systematic work and to evolve a plan for the conservation of freshwater fish biodiversity.

MATERIALS AND METHODS

The present checklist is based on the available published literature present in the form of research articles, monographs, books, species checklists and technical reports. Jayaram (2010), Fishbase (Froese and Pauly, 2013) and Catalog of fishes by Eschmeyer (Eschmeyer *et al.*, 2016) have been referred for confirmation and updating the orders, families, genera and species. Information on the environment, endemic, exotic status mazimum size and human use (existence or nonexistence of fishery, aquaculture, ornamental or gamefish) of all species was obtained by retrieving species level information from the published literature as well as FishBase (Froese and Pauly, 2013). The endemicity or exoticity of all species indicated based on Indian subcontinent. The list of ornamental fishes is prepared based on colouration pattern, shape and maximum size. Information on the conservation status of all taxa in this paper was retrieved from the International Union for Conservation of Nature (IUCN) red list of threatened species, downloaded on 27 January 2016 (IUCN, 2015).

RESULTS

The present checklist revealed occurrence of 227 species of fishes belonging to 108 genera, 37 families and 12 orders from the freshwater bodies of United Andhra Pradesh (Table 2). These included 136 primary freshwater species, 74 freshwater cum brackish species and 17 marine cum freshwater cum brackish species. Freshwater fishes of United Andhra Pradesh consist of 79 species with ornamental value, 51 species forms candidate species for aquaculture, 134 species are important for capture fishery and 25 species are important for game fishing or sport fishing (Table 1). The top three orders with diverse species composition were Cypriniformes (122 species, 45 genera and 3 families), Siluriformes (42 species, 20 genera and 9 families) and Perciformes (34 species, 21 genera and 11 families) (Table 2). The most diverse family was Cyprinidae with a total of 111 species from 40 genera (48.89 %) followed by Bagridae with 17 species and 4 genera and Schilbeidae with 9 species and 6 genera (Table 2).

Of the 227 species, 16 species are exotic and alien, 5 species are endemic to United Andhra Pradesh, 2 species are endemic to Maharashtra, 18 species are endemic to Western Ghats of India, 7 species are endemic to Peninsular India and 42 species are endemic to India. *Barilius evezardi* is endemic to Maharashtra and *Neotropius khavalchor* is endemic to Maharashtra and United Andhra Pradesh (Table 1).

The freshwater fish fauna of this state has been reviewed in respect of the threatened fishes and this has revealed that United Andhra Pradesh contains 20 threatened freshwater fishes (3 critically endangered species, 11 endangered species and 6 vulnerable species) as per IUCN conservation status. There are 207 species were under the non-threatened category, among which 14 species were nearly threatened, 164 species were least concern, 12 species were data deficient whereas 17 species have not been evaluated for IUCN status (Table 1).

Species such as *Puntius thomassi*, *Oreochromis variabilis* and *Hemibagrus punctatus* are critically endangered. The endangered category includes *Hypselobarbus curmuca*, *Labeo potail*, *Puntius fraseri*, *Schismatorhynchos nukta*, *Thynnichthys sandkhol*, *Tor khudree*, *Hypselobarbus mussullah*, *Longischistura striata*, *Etroplus canarensis*, *Pangasianodon hypophthalmus* and *Silonia childreni*. The species under Vulnerable category are *Cirrhinus cirrhosus*, *Cyprinus carpio*, *Hypselobarbus kolus*, *Laubuka fasciata*, *Salmophasia horai* and *Gagata itchkeea* (Table 1).

DISCUSSION

In this paper a checklist of freshwater fishes of United Andhra Pradesh is presented along with their environment, endemic or exotic status, maximum size human use and conservation status (Table 1). It is observed that the checklist comprise mainly freshwater fishes and estuarine fishes, chiefly constituted by the orders cypriniformes (53.74%), siluriformes (18.50%) and perciformes (14.97%). Similar dominance of cypriniformes, siluriformes and perciformes were indicated by Devi and Indra (2003); Chandrasekhar (2004); Barman (2009); Rao *et al.* (2013) and Thirupathaiah *et al.* (2013). As numbers of commercial, sports, ornamental and cultivable fishes are high, commercial and recreational fishing could be organized and seed production by selective breeding is recommended for culture of these species.

Present record of 227 species is relatively higher compared to all the following reports. Barman (1993) reported the occurrence of 158 species in 68 genera belonging to 27 families of 10 orders in Andhra Pradesh. Chandrasekhar (2004) reported 65 species of freshwater fishes belonging to 36 genera belonging to 13 families of five orders from Hyderabad and its vicinity. Barman (2009) recorded 61 species of freshwater fishes from United Andhra Pradesh. Rao *et al.* (2013) listed 63 species from 9 orders, 22 families and 41 genera from River Champavathi in Vizianagaram District of United Andhra Pradesh. Ramulu and Benarjee (2013) listed 30 species belonging to 13 families and 6 orders from Nagaram Tank in Warangal District of Telangana. Reddy *et al.* (2013) reported 30 species of freshwater fishes belonging to 10 families and 21 genera from Thummalapalle Uranium Mining Area of United Andhra Pradesh. Thirupathaiah *et al.* (2013) recorded 44 species from 8 orders, 16 families and 26 genera from Lower Manair Reservoir in Karimnagar District of United Andhra Pradesh. Raju *et al.* (2014) reported 92 species of fish belonging to 13 orders, 34 families and 57 genera from Lake Kolleru of United Andhra Pradesh. Rao (2014a) recorded fifty three ornamental fish species belong to eight orders, 19 families and 34 genera from Lower Manair Dam at Karimnagar District of United Andhra Pradesh. Rao (2014b) recorded sixty four fish species belong to eight orders, 19 families and 39 genera from Lower Manair Dam at Karimnagar District of United Andhra Pradesh. Thirupathaiah *et al.* (2014) recorded 25 fishes belonging to 7 orders, 11 families and 18 genera from Kamalapur lake in Karimnagar District of Telangana. Hemanthkumar *et al.* (2015) recorded 41 species belong to 6 orders, 12 families and 27 genera from River Krishna at Vijayawada Region of United Andhra Pradesh. Laxmappa *et al.* (2015) recorded 109 fish species belonging to 7 orders 19 families and 46 genera from Krishna River in Mahabubnagar district of Telangana.

REFERENCES

Aruna, M., Vankara, A.P. & Gudivada, M. 2011. A Report on the Occurrence of Some Endoparasitic Helminths in Selected Fish Species of Tenali, Guntur District, Andhra Pradesh, India. The Bioscan, 6(3): 501-504.

Barman, R.P. 1993. Pisces: Freshwater Fishes, pp.89-334. In: State Fauna Series 5, Fauna of Andhra Pradesh, Part-I. Zoological Survey of India.

Barman, R.P. 2009. Freshwater fish fauna of Andhra Pradesh with comments on the threatened and endemic species. Records of Zoological Survey of India, 109 (Part-l): 41-47.

Chacko, P.I. 1949. The Krishna River and its fishes. Proceedings of the 36th Indian Science Congress 3: 165-165.

Chandrasekhar, S.V.A. 2004. Fish Fauna of Hyderabad and its Environs. Zoos' Print Journal, 19(7): 1530-1533.

David, A. 1963. Studies on fish and fisheries of the Godavari and the Krishna river systems. Part 1, Proceedings of the National Academy of Science, Section B, 33(2): 263-286.

Devi, K.R., and Indra, T.J. 2003. An updated checklist of Ichthyofauna of Eastern Ghats. Zoos' Print Journal, 18(4): 1067-1070.

Devi, K.R., Indra, T.J., Raghunathan, M.B. and Srivastava, O.P. 2005. On some additional records of fish from Andhra Pradesh, India. Records of Zoological Survey of India, 105 (3-4): 21-28.

Eschmeyer, W. N., Fricke, R. and van der Laan, R. (eds). 2016. Catalog of fishes: genera, species, references. Electronic version accessed 19-28 March 2016. (http://researcharchive.calacademy.org/research/ichthyology/catalog/fishcatmain.asp). [This version was edited by Bill Eschmeyer.]

Froese, R. and Pauly, D. (Eds.) 2013. List of Freshwater Fishes reported from India. In: FishBase. www.fishbase.org, 23 October 2015.

IUCN, 2015. The IUCN Red List of Threatened Species. Version2015.2. IUCN, Gland, Switzerland and Cambridge, UK, http://www.iucnredlist.org. 23 October 2015.

*Hemanthkumar, V., Krishna, P.V. &.Rao, K.M. 2015. Fish faunal Diversity and Conservation Status of River Krishna at Vijayawada Region, Andhra Pradesh, India. International Journal of Advanced Research. 3(8): 1040 – 1045.

Jayaram, K.C. 2010. The Freshwater Fishes of the Indian Region. 2nd Edition, Narendra Publishing House, New Delhi, 616 pp.

Jerdon, T.C. 1849. On the freshwater fishes of southern India. Madras Journal of Lit. Science 15: 302-346.

*Laxmappa, B., Bakshi, R.R. & Narayana, D.V.S. 2015. Studies on ichthyofaunal diversity of Krishna River in Mahabubnagar district, Telangana, India. International Journal of Fisheries and Aquatic Studies, 2(5): 99-104.

Mahmood, S. and Rahimullah (1947). Fish survey of Hyderabad state. Part IV. Fishes of Nizamabad district. Journal of the Bombay Natural History Society 47: 102-111.

Misra, K.S. 1938. On a collection of fish from the Eastern Ghats. Records of the Indian Museum 40(3): 255-264.

Murthy, V.S. 1977. Taxonomic studies on the fishes of the family Cyprinidae from lake Kolleru, Andhra Pradesh. Proceedings of the Indian Academy of Science 85 B (3): 107-146.

Rahimullah, M. 1943. Fish Survey of Hyderabad State, Part-II. Fishes of Hyderabad city and its suburbs. Journal of the Bombay Natural History Society 44(1&2): 88-95.

Rahimullah, M. 1944. Fish survey of Hyderabad state III. Journal of the Bombay Natural History Society 45: 73-77.

*Raju, C.S., Rao, J.C.S. & Simhachalam, G. 2014. Biodiversity and Conservation status of Ichthyofauna of Lake Kolleru, Andhra Pradesh, India. International Journal of Scientific Research, 3(5): 555-563.

Ramulu, K.N. & Benarjee, G. 2013. Fish Species Diversity of Nagaram Tank of Warangal, Andhra Pradesh. Journal of Environmental Science, Toxicology and Food Technology, 3(4): 14-18.

Rao, C.A.N., Deepa, J. & Hakeel, Md. 2011. Comparative account on icthyofauna of Pocharam and Wyra lakes of Andhra Pradesh, India. Journal of Threatened Taxa, 3(2): 1564-1566.

Rao, J.C., Rao, G.K., Raju, Ch.S. & Simhachalam G. 2015. Larvicidal efficacy of four indigenous ornamental fish species of Lake Kolleru, India. Journal of Biodiversity and Environmental Sciences, 7(1): 164-172.

Rao, J.C.S., Simhachalam, G. and Raju, S.Ch. 2013. A study on Ichthyofaunal diversity, conservation status and anthropogenic stress of river Champavathi, Vizianagaram district (AP) India. Asian J. Exp. Biol. Sci., 4(3): 418-425.

Rao, K.R. 2014a. Diversity of Ornamental Fishes in Lower Manair Dam at Karimnagar Dt. Andhra Pradesh. Journal of Pharmacy and Biological Sciences, 9(1): 20-24.

Rao, K.R. 2014b. Ichthyo faunal bio diversity in the lower Manair Dam at Karimnagar district; Telangana State: India. Advances in Applied Science Research, 5(5): 237-248.

Rao, K.R. 2015. Fish Biomass and Physico-Chemical Characteristics of Lower Manair Reservoir at Karimnagar Dt, Telangana, India. Journal of Applied Science and Research, 3 (4):1-12.

Rao, M. and Reddy, Y.S. 1984. Fish fauna of Hussainsagar, Hyderabad. Jantu 2: 1-16.

Reddy, Y.A., Sadasivaiah, B., Rajakullaiswamy, K., Indira, P. & Pullaiah, T. 2013.

Ichthyofauna of Thummalapalle Uranium Mining Area, Andhra Pradesh, India. World Journal of Zoology, 8 (1): 62-66.

Srikanth, K., Ramu, G. & Benarjee, G. (2009). The study on fish diversity of Rammappa lake Warangal District, Andhra Pradesh, India. Aquatic Biology, Pp 1-9.

Sugunan, V.V., 1995. Reservoir fisheries of India. Technical paper no. 345, FAO of U.N., Rome. 423p.

Sykes, W.H. 1838. On the fishes of Deccan. Proceedings of the Zoological Society, London, 6: 157-165.

Thirupathaiah, M., Samatha, C. & Sammaiah, Ch. (2014). Diversity and Conservation Status of Fish Fauna in Freshwater Lake of Kamalapur, Karimnagar District, Telangana, India. Journal of Environmental Science, Toxicology and Food Technology. 8(5): 19-24.

Thirupathaiah, M., Samatha, C.H., & Sammaiah, C. H. 2013. A Checklist of Freshwater Fishes of the Lower Manair Reservoir in Karimnagar District, AP, India. Research Journal of Animal, Veterinary and Fishery Sciences, 1(6), 10-14.

* Referred from predatory journals (Further confirmation is required for the species reported from predatory journals)

Table 1. Freshwater Fishes of United Andhra Pradesh with note on environment, endemic or exotic status, human use and conservation status

Order, Family, Scientific name	Environment	Endemic, Exotic	Maximum size (TL)	Human Use	IUCN Status	References
Order: Anguilliformes						
Family: Anguillidae						
Anguilla bengalensis (Gray, 1831)	Marine; freshwater; brackish		200 cm	Capture Fishery; Aquaculture; Gamefish	NT	Devi & Indra (2003); Barman (2009); Rao *et al.* (2013); Rao (2014a); Rao (2014b)
Anguilla bicolor McClelland, 1844	Marine; freshwater; brackish		123 cm	Capture Fishery	NT	Rao *et al.* (2013); Thirupathaiah *et al.* (2013); Rao (2014a); Rao (2014b); Thirupathaiah *et al.* (2014)
Anguilla nebulosa McClelland, 1844	Marine; freshwater; brackish		121 cm	Capture Fishery	LC	Rao *et al.* (2013)
Family: Moringuidae						
Moringua raitaborua (Hamilton, 1822)	Freshwater; brackish		44.0 cm SL	Fisheries: of no interest	NE	Rao *et al.* (2013)
Order: Beloniformes						
Family: Adrianichthyidae						
Oryzias dancena (Hamilton, 1822)	Freshwater; brackish		3.1 cm SL	Fisheries: of no interest	LC	Rao *et al.* (2013)
Oryzias melastigma (McClelland, 1839)	Freshwater; brackish		4.0 cm	Fisheries: of no interest	LC	Devi & Indra (2003)

Order, Family, Scientific name	Environment	Endemic, Exotic	Maximum size (TL)	Human Use	IUCN Status	References
Family: Belonidae						
Xenentodon cancila (Hamilton, 1822)	Marine; freshwater; brackish		40.0 cm	Capture Fishery; Ornamental	LC	Devi & Indra (2003); Rao *et al.* (2011); Rao *et al.* (2013); Thirupathaiah *et al.* (2013); Thirupathaiah *et al.* (2014)
Family: Hemiramphidae						
Hyporhamphus limbatus (Valenciennes, 1847)	Marine; freshwater; brackish		35.0 cm	Capture Fishery	LC	Laxmappa *et al.* (2015)
Hyporhamphus quoyi (Valenciennes, 1847)	Marine; freshwater; brackish		31.2 cm SL	Capture Fishery	NE	Thirupathaiah *et al.* (2013); Rao (2014a); Rao (2014b); Thirupathaiah *et al.* (2014); Rao (2015)
Rhynchorhamphus georgii (Valenciennes, 1847)	Marine; freshwater; brackish		31.0 cm	Capture Fishery	NE	Laxmappa *et al.* (2015)
Order: Characiformes						
Family: Serrasalmidae						
Pygocentrus nattereri Kner, 1858	Freshwater	EX	50.0 cm SL	Capture Fishery; Ornamental	NE	Raju *et al.* (2014)
Order: Clupeiformes						
Family: Clupeidae						
Ehirava fluviatilis Deraniyagala, 1929	Marine; freshwater; brackish		5.0 cm SL	Capture Fishery	NE	Devi *et al.* (2005)
Gonialosa manmina (Hamilton, 1822)	Freshwater; brackish		14.1 cm	Capture Fishery	LC	Devi & Indra (2003)

Order, Family, Scientific name	Environment	Endemic, Exotic	Maximum size (TL)	Human Use	IUCN Status	References
Gudusia chapra (Hamilton, 1822)	Freshwater; brackish		20.0 cm	Capture Fishery	LC	Devi & Indra (2003)
Tenualosa ilisha (Hamilton, 1822)	Marine; freshwater; brackish		42 cm	Capture Fishery; Aquaculture	LC	Barman (2009)
Order: Cypriniformes						
Family: Cobitidae						
Lepidocephalichthys berdmorei (Blyth, 1860)	Freshwater		8.0 cm SL	Fisheries: of no interest	LC	Rao (2014a); Rao (2014b)
Lepidocephalichthys coromandelensis (Menon, 1992)	Freshwater	ENAP	5.7 cm SL	Fisheries: of no interest	LC	Devi & Indra (2003)
Lepidocephalichthys guntea (Hamilton, 1822)	Freshwater; brackish		15.0 cm	Ornamental	LC	Devi & Indra (2003); Chandrasekhar (2004); Srikanth *et al.* (2009); Ramulu & Benarjee (2013); Rao (2014b); Rao (2014a)
Lepidocephalichthys thermalis (Valenciennes, 1846)	Freshwater		38.0 cm SL	Ornamental	LC	Devi & Indra (2003); Reddy *et al.* (2013)
Family: Cyprinidae						
Amblypharyngodon microlepis (Bleeker, 1853)	Freshwater	ENI	10.0 cm	Ornamental	LC	Devi & Indra (2003); Srikanth *et al.* (2009); Thirupathaiah *et al.* (2013); Rao (2014a); Rao (2014b); Thirupathaiah *et al.* (2014)
Amblypharyngodon mola (Hamilton, 1822)	Freshwater		20.0 cm	Fisheries: of no interest	LC	Devi & Indra (2003); Chandrasekhar (2004); Ramulu & Benarjee (2013); Rao *et al.* (2013); Thirupathaiah *et al.* (2013); Rao (2014a); Thirupathaiah *et al.* (2014); Rao *et al.* (2015)

Order, Family, Scientific name	Environment	Endemic, Exotic	Maximum size (TL)	Human Use	IUCN Status	References
Barilius bakeri Day, 1865	Freshwater	ENWG	15.0 cm	Ornamental	LC	Laxmappa *et al.* (2015)
Barilius barila (Hamilton, 1822)	Freshwater		10.0 cm	Fisheries: of no interest	LC	Devi & Indra (2003); Barman (2009)
Barilius barna (Hamilton, 1822)	Freshwater		15.0 cm	Capture Fishery	LC	Devi & Indra (2003); Chandrasekhar (2004)
Barilius bendelisis (Hamilton, 1807)	Freshwater		22.7 cm	Capture Fishery	LC	Devi & Indra (2003); Chandrasekhar (2004); Hemanthkumar *et al.* (2015)
Barilius evezardi Day, 1872	Freshwater	ENMH	11.0 cm	Ornamental	DD	Barman (2009)
Cabdio morar (Hamilton, 1822)	Freshwater		20.0 cm	Fisheries: of no interest	LC	Devi & Indra (2003)
Carassius carassius (Linnaeus, 1758)	Freshwater; brackish	EX	64.0 cm	Capture Fishery; Aquaculture; Gamefish; Ornamental	LC	Devi & Indra (2003)
Catla catla (Hamilton, 1822)	Freshwater; brackish	ENI	182 cm	Capture Fishery; Aquaculture; Gamefish	LC	Devi & Indra (2003); Chandrasekhar (2004); Aruna *et al.* (2011); Rao *et al.* (2013); Thirupathaiah *et al.* (2013); Thirupathaiah *et al.* (2014)
Chagunius chagunio (Hamilton, 1822)	Freshwater		50.0 cm	Capture Fishery; Gamefish	LC	Devi & Indra (2003);
Chela cachius (Hamilton, 1822)	Freshwater; brackish		6.0 cm	Ornamental	LC	Devi & Indra (2003); Chandrasekhar (2004); Rao *et al.* (2013)
Cirrhinus cirrhosus (Bloch, 1795)	Freshwater; brackish	ENI	100.0 cm SL	Capture Fishery; Aquaculture; Gamefish	VU	Barman (2009)

Order, Family, Scientific name	Environment	Endemic, Exotic	Maximum size (TL)	Human Use	IUCN Status	References
Cirrhinus fulungee (Sykes, 1839)	Freshwater	ENPI	30.0 cm SL	Capture Fishery	LC	Devi & Indra (2003); Barman (2009)
Cirrhinus mrigala (Hamilton, 1822)	Freshwater	ENI	99.0 cm	Fisheries: of no interest	LC	Devi & Indra (2003); Chandrasekhar (2004); Aruna *et al.* (2011); Rao *et al.* (2013); Thirupathaiah *et al.* (2013); Thirupathaiah *et al.* (2014)
Cirrhinus reba (Hamilton, 1822)	Freshwater		30.0 cm	Fisheries: of no interest	LC	Devi & Indra (2003); Chandrasekhar (2004); Rao *et al.* (2013); Thirupathaiah *et al.* (2013); Rao (2014a); Rao (2014b); Thirupathaiah *et al.* (2014)
Crossocheilus latius (Hamilton, 1822)	Freshwater; brackish	ENI	15.2 cm	Fisheries: of no interest	LC	Devi & Indra (2003)
Ctenopharyngodon idella (Valenciennes, 1844)	Freshwater	EX	150 cm	Capture Fishery; Aquaculture; Gamefish	NE	Rao *et al.* (2013); Rao (2014b); Rao (2015)
Cyprinus carpio Linnaeus, 1758	Freshwater; brackish	EX	120 cm	Capture Fishery; Aquaculture; Gamefish; Ornamental	VU	Chandrasekhar (2004); Srikanth *et al.* (2009); Ramulu & Benarjee (2013); Rao *et al.* (2013); Thirupathaiah *et al.* (2013); Rao (2014a); Rao (2014b); Thirupathaiah *et al.* (2014); Rao (2015)
Danio rerio (Hamilton, 1822)	Freshwater		3.8 cm SL	Fisheries: of no interest; Ornamental	LC	Devi & Indra (2003); Chandrasekhar (2004)
Devario aequipinnatus (McClelland, 1839)	Freshwater		15.0 cm	Ornamental	LC	Devi & Indra (2003); Reddy *et al.* (2013)
Devario devario (Hamilton, 1822)	Freshwater		10.0 cm	Ornamental	LC	Devi & Indra (2003); Rao *et al.* (2013); Rao (2014a); Rao (2015); Rao (2014b)
Devario malabaricus (Jerdon, 1849)	Freshwater		12.0 cm	Ornamental	LC	Devi & Indra (2003)

Order, Family, Scientific name	Environment	Endemic, Exotic	Maximum size (TL)	Human Use	IUCN Status	References
Esomus barbatus (Jerdon, 1849)	Freshwater	ENI	12.0 cm	Ornamental	LC	Devi & Indra (2003)
Esomus danricus (Hamilton, 1822)	Freshwater; brackish		13.0 cm	Capture Fishery; Ornamental	LC	Devi & Indra (2003); Chandrasekhar (2004); Srikanth *et al.* (2009); Rao *et al.* (2013); Rao *et al.* (2015)
Esomus thermoicos (Valenciennes, 1842)	Freshwater		12.7 cm	Capture Fishery; Ornamental	LC	Devi & Indra (2003); Devi *et al.* (2005)
Garra gotyla gotyla (Gray, 1830)	Freshwater		18.0 cm	Capture Fishery	LC	Barman (2009); Rao (2014a); Rao (2014b); Rao (2015);
Garra gotyla stenorhynchus Jerdon, 1849	Freshwater	ENWG	15.5 cm SL	Fisheries: of no interest	LC	Chandrasekhar (2004); Barman (2009)
Garra lamta (Hamilton, 1822)	Freshwater		20.4 cm	Capture Fishery	LC	Reddy *et al.* (2013)
Garra mcclellandi (Jerdon, 1849)	Freshwater	ENI	17.4 cm SL	Fisheries: of no interest	LC	Barman (2009)
Garra mullya (Sykes, 1839)	Freshwater	ENI	17.0 cm	Fisheries: of no interest	LC	Devi & Indra (2003); Chandrasekhar (2004); Reddy *et al.* (2013)
Hypophthalmichthys molitrix (Valenciennes, 1844)	Freshwater	EX	105 cm	Capture Fishery; Aquaculture	NT	Devi & Indra (2003)
Hypophthalmichthys nobilis (Richardson, 1845)	Freshwater	EX	146 cm SL	Capture Fishery; Aquaculture; Ornamental	DD	Devi *et al.* (2005)
Hypselobarbus curmuca (Hamilton, 1807)	Freshwater	ENI	120 cm	Capture Fishery; Aquaculture	EN	Barman (2009)
Hypselobarbus dobsoni (Day, 1876)	Freshwater	ENWG	120 cm	Aquaculture	DD	Laxmappa *et al.* (2015)
Hypselobarbus jerdoni (Day, 1870)	Freshwater	ENWG	46.0 cm	Capture Fishery	LC	Barman (2009)

Order, Family, Scientific name	Environment	Endemic, Exotic	Maximum size (TL)	Human Use	IUCN Status	References
Hypselobarbus kolus (Sykes, 1839)	Freshwater	ENI	30.0 cm	Fisheries: of no interest	VU	Chandrasekhar (2004)
Hypselobarbus lithopidos (Day, 1874)	Freshwater	ENI	60.0 cm	Capture Fishery	DD	Hemanthkumar *et al.* (2015)
Bangana ariza (Hamilton, 1807)	Freshwater	ENI	30.0 cm	Capture Fishery	LC	Devi & Indra (2003); Barman (2009);Thirupathaiah *et al.* (2013); Rao (2014a); Rao (2014b); Rao (2015)
Labeo angra (Hamilton, 1822)	Freshwater		22.0 cm	Capture Fishery	LC	Devi & Indra (2003);
Labeo bata (Hamilton, 1822)	Freshwater		61.0 cm	Capture Fishery; Aquaculture	LC	Devi & Indra (2003); Rao *et al.* (2013); Rao (2014a); Rao (2014b); Rao (2015)
Labeo boga (Hamilton, 1822)	Freshwater		30.0 cm	Fisheries: of no interest	LC	Laxmappa *et al.* (2015)
Labeo boggut (Sykes, 1839)	Freshwater		29.0 cm	Capture Fishery; Aquaculture	LC	Devi & Indra (2003); Chandrasekhar (2004); Reddy *et al.* (2013)
Labeo caeruleus Day, 1877	Freshwater		35.0 cm	Capture Fishery	NE	Laxmappa *et al.* (2015)
Labeo calbasu (Hamilton, 1822)	Freshwater; brackish		90.0 cm	Capture Fishery; Aquaculture	LC	Devi & Indra (2003); Chandrasekhar (2004); Srikanth *et al.* (2009); Rao *et al.* (2013); Thirupathaiah *et al.* (2013); Rao (2014a); Rao (2014b); Ramulu & Benarjee (2013); Rao (2015)
Labeo dussumieri (Valenciennes, 1842)	Freshwater		50.0 cm	Capture Fishery; Aquaculture	LC	Barman (2009)
Labeo dyocheilus (McClelland, 1839)	Freshwater		90.0 cm	Capture Fishery	LC	Laxmappa *et al.* (2015)

Order, Family, Scientific name	Environment	Endemic, Exotic	Maximum size (TL)	Human Use	IUCN Status	References
Labeo fimbriatus (Bloch, 1795)	Freshwater		91.0 cm	Capture Fishery; Aquaculture	LC	Devi & Indra (2003); Chandrasekhar (2004); Rao *et al.* (2013); Thirupathaiah *et al.* (2013); Rao (2014a); Rao (2014b); Rao (2015)
Labeo gonius (Hamilton, 1822)	Freshwater		150 cm	Capture Fishery; Aquaculture	LC	Devi & Indra (2003)
Labeo kawrus (Sykes, 1839)	Freshwater	ENWG	60.0 cm	Capture Fishery; Aquaculture	LC	Barman (2009)
Labeo kontius (Jerdon, 1849)	Freshwater	ENI	61.0 cm	Capture Fishery; Aquaculture	LC	Laxmappa *et al.* (2015)
Labeo microphthalmus Day, 1877	Freshwater		25.0 cm	Capture Fishery	LC	Laxmappa *et al.* (2015)
Labeo pangusia (Hamilton, 1822)	Freshwater		90.0 cm	Capture Fishery	LC	Barman (2009); Ramulu & Benarjee (2013); Rao *et al.* (2013)
Labeo porcellus (Heckel, 1844)	Freshwater		35.0 cm	Capture Fishery	LC	Barman (2009) ; Rao (2014b); Rao (2015)
Labeo potail (Sykes, 1839)	Freshwater	ENI	30.0 cm	Capture Fishery	EN	Chandrasekhar (2004); Barman (2009); Srikanth *et al.* (2009); Ramulu & Benarjee (2013)
Labeo rohita (Hamilton, 1822)	Freshwater; brackish		200 cm	Capture Fishery; Aquaculture; Gamefish	LC	Devi & Indra (2003); Chandrasekhar (2004); Aruna *et al.* (2011); Rao *et al.* (2013); Thirupathaiah *et al.* (2013)
Laubuka fasciata (Silas, 1958)	Freshwater		6.0 cm	Fisheries: of no interest	VU	Hemanthkumar *et al.* (2015)
Laubuka laubuca (Hamilton, 1822)	Freshwater; brackish		7.0 cm	Ornamental	LC	Devi & Indra (2003); Rao *et al.* (2011); Rao *et al.* (2013)
Megarasbora elanga (Hamilton, 1822)	Freshwater		21.0 cm	Fisheries: of no interest	LC	Srikanth *et al.* (2009); Thirupathaiah *et al.* (2013); Rao (2014a); Rao (2014b)

Order, Family, Scientific name	Environment	Endemic, Exotic	Maximum size (TL)	Human Use	IUCN Status	References
Oreichthys cosuatis (Hamilton, 1822)	Freshwater		8.0 cm	Fisheries: of no interest	LC	Devi & Indra (2003);
Osteobrama belangeri (Valenciennes, 1844)	Freshwater		38.0 cm SL	Capture Fishery; Aquaculture	NT	Chandrasekhar (2004)
Osteobrama cotio (Hamilton, 1822)	Freshwater		15.0 cm	Fisheries: of no interest	LC	Devi & Indra (2003); Chandrasekhar (2004); Rao *et al.* (2013); Thirupathaiah *et al.* (2013); Rao (2014a); Rao (2015)
Osteobrama cunma (Day, 1888)	Freshwater		15.0 cm	Fisheries: of no interest	LC	Devi & Indra (2003); Barman (2009)
Osteobrama neilli (Day, 1873)	Freshwater	ENI	12.0 cm	Fisheries: of no interest	LC	Devi & Indra (2003); Barman (2009)
Osteobrama peninsularis Silas, 1952	Freshwater	ENPI	15.0 cm	Capture Fishery	DD	Rao *et al.* (2011)
Osteobrama vigorsii (Sykes, 1839)	Freshwater; brackish	ENI	30.0 cm	Capture Fishery	LC	Devi & Indra (2003); Chandrasekhar (2004); Barman (2009); Rao *et al.* (2011)
Osteochilus nashii (Day, 1869)	Freshwater	ENI	18.0 cm	Capture Fishery	LC	Barman (2009)
Osteochilus vittatus (Valenciennes, 1842)	Freshwater	EX	32.0 cm SL	Capture Fishery; Aquaculture; Ornamental	LC	Laxmappa *et al.* (2015)
Osteochilichthys brevidorsalis (Day, 1873)	Freshwater	ENI	15.0 cm	Fisheries: of no interest	LC	Hemanthkumar *et al.* (2015)
Osteochilichthys thomassi (Day, 1877)	Freshwater	ENWG	32.0 cm	Capture Fishery; Aquaculture	LC	Barman (2009)
Parapsilorhynchus tentaculatus (Annandale, 1919)	Freshwater	ENI	4.5 cm	Fisheries: of no interest	LC	Devi & Indra (2003);

Order, Family, Scientific name	Environment	Endemic, Exotic	Maximum size (TL)	Human Use	IUCN Status	References
Barbodes carnaticus (Jerdon, 1849)	Freshwater	ENI	60.0 cm	Capture Fishery; Aquaculture	LC	Laxmappa *et al.* (2015)
Dawkinsia filamentosa (Valenciennes, 1844)	Freshwater; brackish	ENI	18.0 cm	Capture Fishery; Ornamental	LC	Devi & Indra (2003); Chandrasekhar (2004)
Haludaria melanampyx (Day, 1865)	Freshwater	ENWG	7.5 cm	Ornamental	NE	Devi & Indra (2003); Barman (2009)
Pethia conchonius (Hamilton, 1822)	Freshwater		14.0 cm	Fisheries: of no interest; Ornamental	LC	Devi & Indra (2003); Chandrasekhar (2004); Barman (2009)
Pethia gelius (Hamilton, 1822)	Freshwater		5.1 cm	Ornamental	LC	Devi & Indra (2003); Barman (2009); Reddy *et al.* (2013)
Pethia guganio (Hamilton, 1822)	Freshwater		8.0 cm	Fisheries: of no interest	LC	Devi & Indra (2003); Barman (2009)
Pethia phutunio (Hamilton, 1822)	Freshwater		3.5 cm	Ornamental	LC	Devi & Ind ra (2003);
Pethia ticto (Hamilton, 1822)	Freshwater; brackish		10.0 cm	Ornamental	LC	Devi & Indra (2003); Chandrasekhar (2004); Rao *et al.* (2011); Rao *et al.* (2013); Rao (2014a)
Puntius ambassis (Day, 1869)	Freshwater	ENAP	7.5 cm	Fisheries: of no interest	DD	Devi & Indra (2003)
Puntius amphibius (Valenciennes, 1842)	Freshwater; brackish		20.0 cm	Capture Fishery	DD	Devi & Indra (2003); Chandrasekhar (2004); Reddy *et al.* (2013)
Puntius bimaculatus (Bleeker, 1863)	Freshwater		7.0 cm	Ornamental	LC	Devi & Indra (2003); Devi *et al.* (2005)
Puntius chola (Hamilton, 1822)	Freshwater		15.0 cm	Capture Fishery; Ornamental	LC	Devi & Indra (2003); Chandrasekhar (2004); Rao *et al.* (2013); Thirupathaiah *et al.* (2013); Rao (2014a); Thirupathaiah *et al.* (2014)

Order, Family, Scientific name	Environment	Endemic, Exotic	Maximum size (TL)	Human Use	IUCN Status	References
Puntius dorsalis (Jerdon, 1849)	Freshwater		25.0 cm	Capture Fishery; Ornamental	LC	Devi & Indra (2003); Chandrasekhar (2004)
Puntius fraseri (Hora & Misra, 1938)	Freshwater	ENI	4.6 cm	Fisheries: of no interest	EN	Devi *et al.* (2005)
Puntius melanostigma (Day, 1878)	Freshwater	ENI	10.0 cm	Fisheries: of no interest	NE	Chandrasekhar (2004)
Puntius parrah Day, 1865	Freshwater	ENWG	15.0 cm	Capture Fishery	LC	Laxmappa *et al.* (2015)
Puntius sophore (Hamilton, 1822)	Freshwater; brackish		20.0 cm	Ornamental	LC	Devi & Indra (2003); Chandrasekhar (2004); Rao *et al.* (2011);Rao *et al.* (2013); Thirupathaiah *et al.* (2013)
Puntius terio (Hamilton, 1822)	Freshwater		10.0 cm	Ornamental	LC	Devi & Indra (2003); Barman (2009)
Puntius thomassi (Day, 1874)	Freshwater	ENWG	100.0 cm	Capture Fishery	CR	Hemanthkumar *et al.* (2015)
Puntius vittatus Day, 1865	Freshwater; brackish		5.0 cm	Ornamental	LC	Devi & Indra (2003); Barman (2009); Rao *et al.* (2013)
Systomus sarana (Hamilton, 1822)	Freshwater; brackish		42.0 cm	Capture Fishery; Gamefish; Ornamental	LC	Devi & Indra (2003); Chandrasekhar (2004); Barman (2009); Srikanth *et al.* (2009); Reddy *et al.* (2013); Ramulu & Benarjee (2013); Rao *et al.* (2013); Thirupathaiah *et al.* (2013); Rao (2014a); Rao (2015); Thirupathaiah *et al.* (2014)
Raiamas bola (Hamilton, 1822)	Freshwater		35.0 cm	Capture Fishery; Gamefish	LC	Devi & Indra (2003)
Rohtee ogilbii Sykes, 1839	Freshwater	ENI	50.0 cm	Capture Fishery	LC	Barman (2009)
Rasbora caverii (Jerdon, 1849)	Freshwater; brackish		10.0 cm	Capture Fishery; Ornamental	LC	Devi & Indra (2003); Barman (2009); Reddy *et al.* (2013)

Order, Family, Scientific name	Environment	Endemic, Exotic	Maximum size (TL)	Human Use	IUCN Status	References
Rasbora daniconius (Hamilton, 1822)	Freshwater; brackish		15.0	Capture Fishery; Ornamental	LC	Devi & Indra (2003); Chandrasekhar (2004); Rao *et al.* (2011); Reddy *et al.* (2013); Thirupathaiah *et al.* (2013); Rao (2014a); Rao (2014b); Rao *et al.* (2015)
Rasbora labiosa Mukerji, 1935	Freshwater	ENI	8.5 cm	Ornamental	LC	Devi *et al.* (2005)
Rasbora rasbora (Hamilton, 1822)	Freshwater; brackish		13.0 cm	Ornamental	LC	Devi & Indra (2003); Chandrasekhar (2004); Hemanthkumar *et al.* (2015)
Salmophasia acinaces (Valenciennes, 1844)	Freshwater	ENI	15.0 cm	Fisheries: of no interest	LC	Laxmappa *et al.* (2015)
Salmophasia bacaila (Hamilton, 1822)	Freshwater; brackish		18.0 cm	Fisheries: of no interest	LC	Devi & Indra (2003); Chandrasekhar (2004); Rao *et al.* (2011); Ramulu & Benarjee (2013); Thirupathaiah *et al.* (2013); Rao (2014a); Thirupathaiah *et al.* (2014)
Salmophasia balookee (Sykes, 1839)	Freshwater		15.0 cm	Capture Fishery	LC	Devi & Indra (2003); Chandrasekhar (2004); Hemanthkumar *et al.* (2015)
Salmophasia boopis (Day, 1874)	Freshwater	ENI	12.0 cm	Fisheries: of no interest	LC	Barman (2009)
Salmophasia horai (Silas, 1951)	Freshwater	ENI	10.0 cm	Fisheries: of no interest	VU	Barman (2009); Reddy *et al.* (2013)
Salmophasia orissaensis (Banarescu, 1968)	Freshwater	ENI	10.0 cm	Fisheries: of no interest	NE	Devi & Indra (2003)
Salmophasia phulo (Hamilton, 1822)	Freshwater		12.0 cm	Fisheries: of no interest	LC	Thirupathaiah *et al.* (2013); Rao (2014a); Rao (2014b); Thirupathaiah *et al.* (2014); Rao *et al.* (2015); Rao *et al.* (2013)

Order, Family, Scientific name	Environment	Endemic, Exotic	Maximum size (TL)	Human Use	IUCN Status	References
Salmophasia untrahi (Day, 1869)	Freshwater	ENI	20.0 cm	Fisheries: of no interest	LC	Devi & Indra (2003); Chandrasekhar (2004); Barman (2009)
Schismatorhynchos nukta (Sykes, 1839)	Freshwater	ENWG	30.0 cm	Capture Fishery	EN	Barman (2009); Reddy *et al.* (2013)
Securicula gora (Hamilton, 1822)	Freshwater		24.5 cm	Fisheries: of no interest	LC	Devi & Indra (2003)
Thynnichthys sandkhol (Sykes, 1839)	Freshwater	ENPI	46.0 cm	Capture Fishery; Aquaculture	EN	Chandrasekhar (2004); Barman (2009)
Tor khudree (Sykes, 1839)	Freshwater		50.0 cm	Capture Fishery; Aquaculture; Gamefish	EN	Devi & Indra (2003); Chandrasekhar (2004); Barman (2009)
Hypselobarbus mussullah (Sykes, 1839)	Freshwater	ENI	150 cm	Gamefish; Ornamental	EN	Barman (2009)
Family: Nemacheilidae						
Acanthocobitis mooreh (Sykes, 1839)	Freshwater	ENPI	4.4 cm SL	Fisheries: of no interest	LC	Devi & Indra (2003); Barman (2009)
Acanthocobitis botia (Hamilton, 1822)	Freshwater		11.0 cm	Fisheries: of no interest	LC	Devi & Indra (2003)
Indoreonectes evezardi (Day, 1872)	Freshwater	ENWG	3.8 cm SL	Fisheries: of no interest	LC	Devi & Indra (2003)
Longischistura striata (Day, 1867)	Freshwater	ENWG	5.0 cm SL	Fisheries: of no interest	EN	Barman (2009)
Nemacheilus anguilla Annandale, 1919	Freshwater	ENWG	5.0 cm SL	Fisheries: of no interest	LC	Barman (2009)
Nemacheilus corica (Hamilton, 1822)	Freshwater		4.2 cm SL	Fisheries: of no interest	LC	Rao (2014a); Rao (2014b); Rao (2015)

Order, Family, Scientific name	Environment	Endemic, Exotic	Maximum size (TL)	Human Use	IUCN Status	References
Nemacheilus denisoni Day, 1867	Freshwater	ENI	5.0 cm SL	Fisheries: of no interest	LC	Devi & Indra (2003); Rao *et al.* (2011); Barman (2009)
Order: Cyprinodontiformes						
Family: Aplocheilidae						
Aplocheilus panchax (Hamilton, 1822)	Freshwater; brackish		9.0 cm	Capture Fishery; Ornamental	LC	Devi & Indra (2003); Rao *et al.* (2013); Rao *et al.* (2015)
Aplocheilus parvus (Sundara Raj, 1916)	Freshwater; brackish		6.3 cm	Ornamental	NE	Devi *et al.* (2005)
Family: Poeciliidae						
Gambusia affinis (Baird & Girard, 1853)	Freshwater; brackish	EX	4.0 cm	Capture Fishery; Ornamental	LC	Devi & Indra (2003); Chandrasekhar (2004)
Poecilia reticulata Peters, 1859	Freshwater; brackish	EX	5.0 cm SL	Ornamental	NE	Devi & Indra (2003);
Order: Elopiformes						
Family: Megalopidae						
Megalops cyprinoides (Broussonet, 1782)	Marine; freshwater; brackish		45.5 cm SL	Capture Fishery; Aquaculture; Gamefish	DD	Rao *et al.* (2013)
Order: Mugiliformes						
Family: Mugilidae						
Mugil cephalus (Linnaeus, 1758)	Marine; freshwater; brackish		100.0 cm SL	Capture Fishery; Aquaculture; Gamefish	LC	Rao *et al.* (2013)

Order, Family, Scientific name	Environment	Endemic, Exotic	Maximum size (TL)	Human Use	IUCN Status	References
Rhinomugil corsula (Hamilton, 1822)	Freshwater; brackish		45.0 cm	Capture Fishery; Aquaculture	LC	Devi & Indra (2003); Thirupathaiah *et al.* (2013); Rao *et al.* (2011); Thirupathaiah *et al.* (2014); Rao (2014a); Rao (2015); Barman (2009); Rao (2014b)
Order: Osteoglossiformes						
Family: Notopteridae						
Chitala chitala (Hamilton, 1822)	Freshwater		122 cm SL	Capture Fishery; Aquaculture; Gamefish; Ornamental	NT	Devi & Indra (2003); Srikanth *et al.* (2009); Hemanthkumar *et al.* (2015)
Notopterus notopterus (Pallas, 1769)	Freshwater; brackish		60.0 cm SL	Capture Fishery; Aquaculture; Ornamental	LC	Devi & Indra (2003); Chandrasekhar (2004); Rao *et al.* (2011); Rao *et al.* (2013); Thirupathaiah *et al.* (2013)
Order: Perciformes						
Family: Ambassidae						
Chanda nama Hamilton, 1822	Freshwater; brackish		11.0 cm	Ornamental	LC	Devi & Indra (2003); Rao *et al.* (2011); Rao *et al.* (2013); Thirupathaiah *et al.* (2013); Rao (2014a)
Parambassis baculis (Hamilton, 1822)	Freshwater	ENI	5.0 cm SL	Fisheries: of no interest	LC	Devi & Indra (2003)
Parambassis lala (Hamilton, 1822)	Freshwater; brackish		3.9 cm SL	Ornamental	NT	Devi *et al.* (2005)
Parambassis ranga (Hamilton, 1822)	Freshwater; brackish		8.0 cm	Capture Fishery; Ornamental	LC	Devi & Indra (2003); Thirupathaiah *et al.* (2013); Rao (2014a); Thirupathaiah *et al.* (2014); Rao *et al.* (2015)

Order, Family, Scientific name	Environment	Endemic, Exotic	Maximum size (TL)	Human Use	IUCN Status	References
Family: Anabantidae						
Anabas cobojius (Hamilton, 1822)	Freshwater		30.0 cm	Capture Fishery	DD	Aruna *et al.* (2011); Rao *et al.* (2013)
Anabas testudineus (Bloch, 1792)	Freshwater; brackish;		25.0 cm	Capture Fishery; Aquaculture; Ornamental	DD	Devi & Indra (2003); Aruna *et al.* (2011); Ramulu & Benarjee (2013); Rao *et al.* (2013); Rao (2015)
Family: Badidae						
Badis badis (Hamilton, 1822)	Freshwater	ENI	5.0 cm	Ornamental	LC	Devi & Indra (2003); Rao *et al.* (2013)
Family: Channidae						
Channa gachua (Hamilton, 1822)	Freshwater		20.0 cm SL	Ornamental	LC	Rao *et al.* (2013); Rao *et al.* (2015)
Channa marulius (Hamilton, 1822)	Freshwater		183 cm	Capture Fishery; Aquaculture; Gamefish; Ornamental	LC	Devi & Indra (2003); Chandrasekhar (2004); Thirupathaiah *et al.* (2013); Rao (2014a); Rao (2014b); Thirupathaiah *et al.* (2014); Rao (2015)
Channa orientalis Bloch & Schneider, 1801	Freshwater; brackish		33.0 cm	Capture Fishery; Ornamental	NE	Devi & Indra (2003); Chandrasekhar (2004); Barman (2009); Thirupathaiah *et al.* (2013); Ramulu & Benarjee (2013); Rao (2014a); Thirupathaiah *et al.* (2014); Rao (2015)
Channa punctata (Bloch, 1793)	Freshwater; brackish		31.0 cm	Capture Fishery; Aquaculture; Ornamental	LC	Devi & Indra (2003); Chandrasekhar (2004); Aruna *et al.* (2011); Rao *et al.* (2013); Thirupathaiah *et al.* (2013)
Channa striata (Bloch, 1793)	Freshwater; brackish		100.0 cm SL	Capture Fishery; Aquaculture; Ornamental	LC	Devi & Indra (2003); Chandrasekhar (2004); Rao *et al.* (2011); Rao *et al.* (2013); Thirupathaiah *et al.* (2013)

Order, Family, Scientific name	Environment	Endemic, Exotic	Maximum size (TL)	Human Use	IUCN Status	References
Family: Cichlidae						
Etroplus canarensis Day, 1877	Freshwater	ENWG	11.5 cm	Fisheries: of no interest	EN	Raju *et al.* (2014)
Etroplus maculatus (Bloch, 1795)	Freshwater; brackish		8.0 cm	Capture Fishery; Ornamental	LC	Devi & Indra (2003); Chandrasekhar (2004); Rao *et al.* (2013); Rao *et al.* (2011); Rao (2014a); Reddy *et al.* (2013)
Etroplus suratensis (Bloch, 1790)	Freshwater; brackish		40.0 cm	Capture Fishery; Aquaculture; Ornamental	LC	Chandrasekhar (2004); Rao *et al.* (2013); Rao *et al.* (2011); Rao (2014a)
Oreochromis mossambicus (Peters, 1852)	Freshwater; brackish	EX	39.0 cm SL	Capture Fishery; Aquaculture; Gamefish; Ornamental	NT	Devi & Indra (2003); Chandrasekhar (2004); Rao *et al.* (2013); Rao (2014a)
Oreochromis niloticus (Linnaeus, 1758)	Freshwater; brackish	EX	60.0 cm SL	Capture Fishery; Aquaculture	NE	Devi & Indra (2003); Chandrasekhar (2004)
Oreochromis variabilis (Boulenger, 1906)	Freshwater	EX	30.0 cm SL	Capture Fishery; Aquaculture; Ornamental	CR	Devi *et al.* (2005); Rao (2014b)
Family: Gerreidae						
Gerres filamentosus Cuvier, 1829	Marine; freshwater; brackish		35.0 cm	Capture Fishery	LC	Raju *et al.* (2014)

Order, Family, Scientific name	Environment	Endemic, Exotic	Maximum size (TL)	Human Use	IUCN Status	References
Family: Gobiidae						
Awaous grammepomus (Bleeker, 1849)	Freshwater; brackish		15.0 cm SL	Capture Fishery; Ornamental	LC	Rao (2015); Rao (2014b)
Glossogobius giuris (Hamilton, 1822)	Marine; freshwater; brackish		50.0 cm SL	Capture Fishery; Aquaculture; Ornamental	LC	Devi & Indra (2003); Chandrasekhar (2004); Srikanth *et al.* (2009); Rao *et al.* (2011); Rao *et al.* (2013); Ramulu & Benarjee (2013); Thirupathaiah *et al.* (2013); Thirupathaiah *et al.* (2014)
Gobiopsis macrostoma Steindachner, 1861	Marine; freshwater; brackish		10.0 cm SL	Fisheries: of no interest	NE	Rao *et al.* (2013)
Pseudapocryptes elongatus (Cuvier, 1816)	Freshwater; brackish		20.0 cm	Capture Fishery	LC	Raju *et al.* (2014)
Scartelaos histophorus (Valenciennes, 1837)	Marine; freshwater; brackish		14.0 cm SL	Capture Fishery	NE	Raju *et al.* (2014)
Sicyopterus griseus (Day, 1877)	Freshwater; brackish	ENI	9.7 cm	Ornamental	LC	Devi & Indra (2003);
Stigmatogobius sadanundio (Hamilton, 1822)	Freshwater; brackish		9.0 cm SL	Ornamental	NE	Raju *et al.* (2014)
Family: Nandidae						
Nandus nandus (Hamilton, 1822)	Freshwater; brackish		20.0 cm	Capture Fishery; Ornamental	LC	Devi & Indra (2003); Srikanth *et al.* (2009); Ramulu & Benarjee (2013); Rao *et al.* (2013); Rao (2014a); Rao (2014b); Rao (2015)

Order, Family, Scientific name	Environment	Endemic, Exotic	Maximum size (TL)	Human Use	IUCN Status	References
Family: Osphronemidae						
Osphronemus goramy Lacepède, 1801	Freshwater; brackish	EX	70.0 cm SL	Capture Fishery; Aquaculture; Ornamental	LC	Chandrasekhar (2004); Rao *et al.* (2013)
Pseudosphromenus cupanus (Cuvier, 1831)	Freshwater; brackish		7.5 cm	Ornamental	LC	Raju *et al.* (2014)
Trichogaster fasciata Bloch & Schneider, 1801	Freshwater		12.5 cm	Capture Fishery	LC	Chandrasekhar (2004); Rao *et al.* (2013); Thirupathaiah *et al.* (2013); Rao (2014a); Rao (2014b); Thirupathaiah *et al.* (2014); Rao *et al.* (2015)
Trichogaster labiosa Day, 1877	Freshwater	EX	9.0 cm	Ornamental	LC	Rao *et al.* (2015)
Trichogaster lalius (Hamilton, 1822)	Freshwater		8.8 cm	Ornamental	LC	Devi *et al.* (2005); Rao *et al.* (2013); Rao (2014a); Rao (2014b); Rao (2015); Rao *et al.* (2015)
Family: Scatophagidae						
Scatophagus argus (Linnaeus, 1766)	Marine; freshwater; brackish		38.0 cm	Capture Fishery; Aquaculture; Ornamental	LC	Raju *et al.* (2014)
Family: Terapontidae						
Terapon jarbua (Forsskål, 1775)	Marine; freshwater; brackish		36.0 cm	Capture Fishery; Aquaculture	LC	Raju *et al.* (2014)
Order: Siluriformes						
Family: Amblycipitidae						
Amblyceps mangois (Hamilton, 1822)	Freshwater		12.5 cm SL	Fisheries: of no interest	LC	Devi & Indra (2003)

Order, Family, Scientific name	Environment	Endemic, Exotic	Maximum size (TL)	Human Use	IUCN Status	References
Family: Bagridae						
Hemibagrus maydelli (Rössel, 1964)	Freshwater	ENAP	165 cm	Fisheries: of no interest	LC	Laxmappa *et al.* (2015)
Hemibagrus menoda (Hamilton, 1822)	Freshwater	ENI	45.0 cm	Fisheries: of no interest	LC	Devi & Indra (2003)
Hemibagrus microphthalmus (Day, 1877)	Freshwater; brackish		150 cm	Capture Fishery	LC	Laxmappa *et al.* (2015)
Hemibagrus punctatus (Jerdon, 1849)	Freshwater	ENI	45.0 cm	Capture Fishery	CR	Barman (2009)
Mystus armatus (Day, 1865)	Freshwater; brackish		14.5 cm SL	Capture Fishery; Ornamental	LC	Devi & Indra (2003)
Mystus bleekeri (Day, 1877)	Freshwater		15.5 cm	Capture Fishery; Ornamental	LC	Chandrasekhar (2004); Rao *et al.* (2013); Thirupathaiah *et al.* (2013)
Mystus cavasius (Hamilton, 1822)	Freshwater; brackish		40.0 cm SL	Capture Fishery	LC	Devi & Indra (2003); Chandrasekhar (2004); Rao *et al.* (2011); Ramulu & Benarjee (2013); Rao *et al.* (2013); Thirupathaiah *et al.* (2013); Thirupathaiah *et al.* (2014)
Mystus gulio (Hamilton, 1822)	Freshwater; brackish		46.0 cm	Capture Fishery	LC	Devi & Indra (2003); Rao *et al.* (2013); Rao *et al.* (2015)
Mystus malabaricus (Jerdon, 1849)	Freshwater; brackish	ENWG	15.0 cm	Capture Fishery	NT	Barman (2009)
Mystus montanus (Jerdon, 1849)	Freshwater; brackish	ENI	15.0 cm	Capture Fishery	LC	Devi & Indra (2003); Barman (2009)
Mystus tengara (Hamilton, 1822)	Freshwater		18.0 cm	Fisheries: of no interest	LC	Devi & Indra (2003); Thirupathaiah *et al.* (2013); Thirupathaiah *et al.* (2014); Rao (2014a); Rao (2014b)

Order, Family, Scientific name	Environment	Endemic, Exotic	Maximum size (TL)	Human Use	IUCN Status	References
Mystus vittatus (Bloch, 1794)	Freshwater; brackish		21.0 cm SL	Capture Fishery; Ornamental	LC	Devi & Indra (2003); Chandrasekhar (2004); Barman (2009); Rao *et al.* (2011); Rao *et al.* (2013); Thirupathaiah *et al.* (2013)
Rita chrysea Day, 1877	Freshwater	ENPI	19.5 cm	Capture Fishery	LC	Devi & Indra (2003)
Rita gogra (Sykes, 1839)	Freshwater	ENAP	26.0 cm	Capture Fishery	LC	Barman (2009)
Rita kuturnee (Sykes, 1839)	Freshwater	ENPI	30.0 cm	Capture Fishery	LC	Laxmappa *et al.* (2015)
Sperata aor (Hamilton, 1822)	Freshwater		180 cm	Capture Fishery; Gamefish	LC	Devi & Indra (2003); Chandrasekhar (2004); Barman (2009); Rao (2014b)
Sperata seenghala (Sykes, 1839)	Freshwater; brackish		150 cm	Capture Fishery; Aquaculture; Gamefish	LC	Devi & Indra (2003); Chandrasekhar (2004); Barman (2009); Rao *et al.* (2011); Thirupathaiah *et al.* (2014); Rao (2014b)
Family: Clariidae						
Clarias batrachus (Linnaeus, 1758)	Freshwater; brackish		47.0 cm	Capture Fishery; Aquaculture; Ornamental	LC	Devi & Indra (2003); Chandrasekhar (2004); Barman (2009); Rao *et al.* (2013); Thirupathaiah *et al.* (2013); Thirupathaiah *et al.* (2014)
Clarias gariepinus (Burchell, 1822)	Freshwater	EX	170 cm	Capture Fishery; Aquaculture; Gamefish	LC	Rao *et al.* (2013); Rao (2014b); Rao (2015)
Family: Heteropneustidae						
Heteropneustes fossilis (Bloch, 1794)	Freshwater; brackish		30.0 cm	Capture Fishery; Aquaculture; Ornamental	LC	Devi & Indra (2003); Chandrasekhar (2004); Rao *et al.* (2013); Thirupathaiah *et al.* (2013); Thirupathaiah *et al.* (2014)

Order, Family, Scientific name	Environment	Endemic, Exotic	Maximum size (TL)	Human Use	IUCN Status	References
Family: Erethistidae						
Hara hara (Hamilton, 1822)	Freshwater		13.0 cm	Fisheries: of no interest; Ornamental	LC	Devi & Indra (2003)
Family: Pangasiidae						
Pangasianodon hypophthalmus (Sauvage, 1878)	Freshwater	EX	130 cm SL	Capture Fishery; Aquaculture; Ornamental	EN	Laxmappa *et al.* (2015)
Pangasius pangasius (Hamilton, 1822)	Freshwater; brackish		300 cm SL	Capture Fishery; Aquaculture; Gamefish	LC	Devi & Indra (2003); Barman (2009)
Family: Schilbeidae						
Ailia coila (Hamilton, 1822)	Freshwater; brackish		30.0 cm	Capture Fishery	NT	Devi & Indra (2003); Barman (2009)
Clupisoma garua (Hamilton, 1822)	Freshwater; brackish		60.9 cm SL	Capture Fishery; Gamefish	LC	Devi & Indra (2003); Hemanthkumar *et al.* (2015)
Eutropiichthys goongwaree (Sykes, 1839)	Freshwater	ENPI	64.0 cm	Fisheries: of no interest	DD	Laxmappa *et al.* (2015)
Eutropiichthys murius (Hamilton, 1822)	Freshwater		28.0 cm	Capture Fishery	LC	Devi & Indra (2003)
Eutropiichthys vacha (Hamilton, 1822)	Freshwater; brackish		34.0 cm	Capture Fishery; Gamefish	LC	Devi & Indra (2003); Thirupathaiah *et al.* (2013); Rao (2014a); Rao (2014b); Thirupathaiah *et al.* (2014)
Neotropius atherinoides (Bloch, 1794)	Freshwater; brackish		15.0 cm	Capture Fishery; Ornamental	LC	Devi & Indra (2003); Barman (2009); Rao *et al.* (2013); Rao (2014b)

Order, Family, Scientific name	Environment	Endemic, Exotic	Maximum size (TL)	Human Use	IUCN Status	References
Neotropius khavalchor Kulkarni, 1952	Freshwater	ENMH, ENAP	15.0 cm	Fisheries: of no interest	DD	Devi & Indra (2003)
Proeutropiichthys taakree (Sykes, 1839)	Freshwater	ENWG	42.0 cm	Capture Fishery	LC	Barman (2009)
Silonia childreni (Sykes, 1839)	Freshwater	ENWG	48.0 cm	Capture Fishery	EN	Barman (2009)
Family: Siluridae						
Ompok bimaculatus (Bloch, 1794)	Freshwater; brackish	ENI	45.0 cm SL	Capture Fishery; Aquaculture; Ornamental	NT	Devi & Indra (2003); Chandrasekhar (2004); Barman (2009); Rao *et al.* (2011); Reddy *et al.* (2013); Rao *et al.* (2013); Thirupathaiah *et al.* (2013); Rao (2014a); Thirupathaiah *et al.* (2014); Hemanthkumar *et al.* (2015)
Ompok malabaricus (Valenciennes, 1840)	Freshwater	ENWG	51.0 cm SL	Aquaculture	LC	Chandrasekhar (2004)
Ompok pabda (Hamilton, 1822)	Freshwater		30.0 cm	Capture Fishery	NT	Devi & Indra (2003); Barman (2009); Rao *et al.* (2013)
Ompok pabo (Hamilton, 1822)	Freshwater		25.0 cm	Capture Fishery	NT	Raju *et al.* (2014)
Wallago attu (Bloch & Schneider, 1801)	Freshwater; brackish		240 cm	Capture Fishery; Gamefish	NT	Devi & Indra (2003); Chandrasekhar (2004); Rao *et al.* (2011); Rao *et al.* (2013); Thirupathaiah *et al.* (2013)
Family: Sisoridae						
Bagarius bagarius (Hamilton, 1822)	Freshwater; brackish		200 cm	Capture Fishery; Gamefish	NT	Devi & Indra (2003); Barman (2009)
Gagata cenia (Hamilton, 1822)	Freshwater; brackish		15.0 cm SL	Capture Fishery	LC	Devi & Indra (2003)

Order, Family, Scientific name	Environment	Endemic, Exotic	Maximum size (TL)	Human Use	IUCN Status	References
Gagata itchkeea (Sykes, 1839)	Freshwater	ENI	7.6 cm	Capture Fishery	VU	Devi & Indra (2003)
Glyptothorax lonah (Sykes, 1839)	Freshwater	ENI	15.0 cm	Fisheries: of no interest	LC	Devi & Indra (2003); Barman (2009)
Order: Synbranchiformes						
Family: Mastacembelidae						
Macrognathus aral (Bloch & Schneider, 1801)	Freshwater; brackish		63.5 cm	Capture Fishery	LC	Devi & Indra (2003); Reddy *et al.* (2013); Rao *et al.* (2013)
Macrognathus guentheri (Day, 1865)	Freshwater	ENI	29.9 cm SL	Capture Fishery; Ornamental	LC	Laxmappa *et al.* (2015)
Macrognathus pancalus Hamilton, 1822	Freshwater; brackish		18.0 cm	Capture Fishery	LC	Devi & Indra (2003); Chandrasekhar (2004); Srikanth *et al.* (2009); Rao *et al.* (2011); Rao *et al.* (2013); Thirupathaiah *et al.* (2014); Rao (2014a)
Mastacembelus armatus (Lacepède, 1800)	Freshwater; brackish		90.0 cm	Capture Fishery; Ornamental	LC	Devi & Indra (2003); Chandrasekhar (2004); Aruna *et al.* (2011); Rao *et al.* (2011); Reddy *et al.* (2013); Rao *et al.* (2013); Thirupathaiah *et al.* (2013)
Family: Synbranchidae						
Monopterus cuchia (Hamilton, 1822)	Freshwater; brackish		70.0 cm	Capture Fishery	LC	Devi & Indra (2003)

ENI = endemic to India; ENPI = endemic to peninsular India; ENWG = endemic to Western Ghats of India; ENMH = endemic to Maharashtra; ENAP = endemic to United Andhra Pradesh; EX – Exotic; TL = Total Length; SL = Standard Length; IUCN - International Union for Conservation of Nature; CR – Critically Endangered; EN – Endangered; VU – Vulnerable; NT – Near Threatened; LC – Least Concern; DD – Data Deficient; NE – Not Evaluated

Table 2. Number of family, genera and species under each Order of Freshwater Fishes of United Andhra Pradesh

Order	Family	Number of Genera	Number of Species
Anguilliformes	Anguillidae	1	3
	Moringuidae	1	1
Beloniformes	Adrianichthyidae	1	2
	Belonidae	1	1
	Hemiramphidae	2	3
Characiformes	Serrasalmidae	1	1
Clupeiformes	Clupeidae	4	4
Cypriniformes	Cobitidae	1	4
	Cyprinidae	40	111
	Nemacheilidae	4	7
Cyprinodontiformes	Aplocheilidae	1	2
	Poeciliidae	2	2
Elopiformes	Megalopidae	1	1
Mugiliformes	Mugilidae	2	2
Osteoglossiformes	Notopteridae	2	2
Perciformes	Ambassidae	2	4
	Anabantidae	1	2
	Badidae	1	1
	Channidae	1	5
	Cichlidae	2	6
	Gerreidae	1	1
	Gobiidae	7	7
	Nandidae	1	1

Order	Family	Number of Genera	Number of Species
	Osphronemidae	3	5
	Scatophagidae	1	1
	Terapontidae	1	1
Siluriformes	Amblycipitidae	1	1
	Bagridae	4	17
	Clariidae	1	2
	Heteropneustidae	1	1
	Erethistidae	1	1
	Pangasiidae	2	2
	Schilbeidae	6	9
	Siluridae	2	5
	Sisoridae	2	4
Synbranchiformes	Mastacembelidae	2	4
	Synbranchidae	1	1
Total = 12	**Total = 37**	**Total = 108**	Total = 227

6

ESTUARINE FISH DIVERSITY OF TAMIL NADU

H. S. Mogalekar, P. Jawahar, P. Pavinkumar, S. Santhoshkumar and C. Sudhan

Department of Fisheries Biology and Resource Management, Fisheries College & Research Institute, Tamil Nadu Fisheries University, Thoothukudi-628 008, Tamil Nadu, India.

INTRODUCTION

Tamil Nadu is the southernmost state in India with coastline of about 1076 km stretches along Bay of Bengal, Arabian Sea and Indian Ocean. The State also has a number of east flowing rivers due to presence of Western Ghats on the west. Cauvery, Vaigai, Tamiraparani, Periyar and Pennar are some of the important rivers of Tamil Nadu which discharges freshwater into Bay of Bengal through the various estuaries (Ramesh *et al.*, 2008). The total estuarine area of Tamil Nadu is estimated to be 56000 ha, which accounts 3.88 % of the total estuarine area of India (De, 2011). The River Cauvery forms the major estuary in Tamil Nadu and the minor estuaries are Araniar, Ennore, Adyar, Cooum, Muttukadu backwaters, Edaiyur-Sadras estuarine complex, Uppanar, Kollidam, Agniar, Kallar, Pinnakayal, Pullavazhi, Athankarai, Kajirangudi, Kottakkarai, Uppar, Vaigai, Kottakkudy, Thengapattanam, Vellar, Pazhayar, Yedayanthittu or Muttukadu, Giriyampeta, Arasalar, Coleroon, Pichavaram, Alangkulam, Sayalkudi, Vaippar, Korampallam Thermal, Manapad and Thopuvizhai, (Jha *et al.*, 2008; Pavinkumar, 2014). Manakudy, Valliyar, and Tengapattinam are the estuaries formed along Arabian Sea in Kanyakumari district of Tamil Nadu (Pavinkumar, 2014). The lagoons are Pulicat lake (South) and Muthupettai (Ramesh *et al.*, 2008).

Estuaries being a dynamic ecosystem provide diverse habitat for the proliferation of diadromous and estuarine resident fish species to complete their life cycle (Jha *et al.*, 2008). They also form a significant component of coastal ecosystem due to their immense biodiversity values in aquatic ecology. The fish fauna inhabiting the estuarine ecosystems of Tamil Nadu are diverse and fairly well known. Significant contributions to systematics of estuarine fish fauna of Tamil Nadu were that of Ramaiyan *et al* (1987); Remadevi *et al.* (2004); Johnson and Selvaraj (2008); Ramanujam and Anbarasan (2008); Kumaran *et al.* (2012); Kannappan and Karthikeyan (2013); Pavinkumar (2014); Ramanujam *et al.* (2014); Sukumaran *et al.* (2014); Bharadhirajan *et al.* (2015); Khan (2015); Mahesh and Saravanakumar (2015); Pavinkumar *et al.* (2015); Raju *et al.* (2015). Since publication of these pioneering works, numerous changes in taxonomy and/or systematic placement of species took place and a comprehensive review on the entire estuarine fish fauna of Tamil Nadu is not available now. The list of species presented earlier by many authors is updated by inclusion of all species of estuarine fishes now known to occur in the Tamil Nadu. Henceforward present checklist is intended to update the scientific nomenclature of all the fish species recorded from estuaries of Tamil Nadu to assist fish taxonomists in future systematic work.

MATERIAL AND METHODS

Estuarine fish fauna of Tamil Nadu was compiled based on publications by Johnson and Selvaraj (2008); Remadevi *et al.* (2004); Ramanujam and Anbarasan (2008); Kumaran *et al.* (2012); Kannappan & Karthikeyan (2013); Ramanujam *et al.* (2014); Sukumaran *et al.* (2014); Bharadhirajan *et al.* (2015); Khan (2015); Mahesh and Saravanakumar (2015); Pavinkumar *et al.* (2015); Raju *et al.* (2015). The records and nomenclature have been updated, listing and classification of estuarine fish species of Tamil Nadu follows Jayaram (2010), Talwar & Kacker (1984), Talwar & Jhingran (1991) and FishBase (Froese & Pauly, 2013). Details on the environment, maximum size [total length (TL) or standard length (SL)] and fisheries status or human use of all species were obtained by retrieving species level information from the published literatures, FishBase (Froese and Pauly, 2013) and with selected field observations. Information on the conservation status of all taxa in this paper was retrieved from the International Union for Conservation of Nature (IUCN) Red List of Threatened Species, whose underlying assessments are based on the IUCN Red List Categories and criteria (Version 2015.2) downloaded on 05 January 2016 (IUCN, 2015).

RESULTS

The comprehensive checklist of estuarine fishes of Tamil Nadu includes a total of 241 different species, which were distributed among 154 genera, 75 families and 17 orders (Table 2). Among these, 99 species were marine cum brackish, 58 were marine cum freshwater cum brackish, 44 were marine, 32 were freshwater cum brackish and the remaining 8 species were freshwater (Table 1). Under the chondrichthyes, three species were recorded representing three orders, three families and three genera (Table 2). Under the osteichthyes, the most diverse order was perciformes which includes 126 species, 77 genera and 38 families. The second most diverse order was the Clupeiformes with 31 species, 15 genera and 5 families (Table 2). The top five families with diverse species composition were carangidae (9 genera, 18 species), cyprinidae (10 genera, 15 species), gobidae (11 genera, 14 species), engraulidae (4 genera, 13 species) and clupeidae (7 genera, 12 species) (Figure 1).

There are 38 estuaries known from Tamil Nadu of which 12 estuaries explored in details for fish biodiversity assessment. One hundred and six species of fishes recorded in Coleroon-Pichavaram-Vellar estuarine complex, 91 species of fishes in Manakudy estuary, 81 species of fishes in Pulicat lake, 73 species of fishes in Yedayanthittu or Muttukadu estuary, 53 species of fishes in Adyar estuary, 33 species of fishes in Giriyampeta estuary and 6 species of fishes in Arasalar estuary (Figure 2).

A consolidated list of estuarine fishes of Tamil Nadu comprises of 27 species with highly commercial value, 110 species with commercial value, 68 species with minor commercial value, 62 species stand for game fishing, 53 species are having ornamental value, 45 species could be used for aquaculture, and 23 species are being used as bait (Table 1).

Based on the information collected on conservation status of all taxa, 3 species listed as vulnerable, 12 as near threatened, 1 species as lower risk-least concern, 65 as least concern, 7 as data deficient and 153 species have not yet been evaluated for their conservation status by IUCN Red list (Table 1).

DISCUSSION

A checklist of the estuarine fishes of Tamil Nadu is presented, along with their classification, distribution in different estuaries of Tamil Nadu, environment in which they occur, fisheries status, human use, maximum size and IUCN Red List conservation status (Table 1). It is observed that the checklist comprise mainly marine fishes and estuarine fishes, chiefly constituted by the orders perciformes (52.28%) and clupeiformes (12.86%). Similar dominance of perciformes and clupeiformes were indicated by Bijukumar and Sushama (2000); Remadevi *et al.* (2004); Ramanujam and Anbarasan (2008); Ghosh *et al.* (2011). As 26 out of 38 estuaries have not been explored for fish biodiversity assessment, many species could be distributed in the unexplored estuarine systems of Tamil Nadu and therefore rigorous biodiversity explorations are obligatory. As numbers of commercial, sports, ornamental and cultivable fishes are high, commercial and recreational fishing could be organized. As 153 species of estuarine fishes of Tamil Nadu has not yet been evaluated for their conservation status by IUCN Red list due to lack of scientific investigations.

Remadevi *et al.* (2004) documented 88 species of which 83 species of estuarine fishes stands valid belonging to 33 families and 11 orders from Pulicat lake. Ramanujam & Anbarasan (2008) listed 75 species under 14 orders and 37 families of which 73 species stands valid from Yedayanthittu or Muttukadu estuary. Ichtyofaunal investigation at Giriyampeta estuary by Kumaran *et al.* (2012) discovered presence of 33 valid species of fishes out of 36 species belonging to 15 families and 21 genera. Fifty three species of fishes stand valid out of 57 species which were recorded in the Adyar estuary by Ramanujam *et al.* (2014). Sukumaran *et al.* (2014) observed 22 species of fishes belonging to 19 genera from the Muthupet estuary. Bharadhirajan *et al.* (2015) described 59 species of finfishes in Coleroon Estuary belonging to 9 Orders, 26 Families and 30 Genera. Ichthyofaunal investigation made by Ramaiyan *et al* (1987) indicated presence of 125 species of fishes from the Vellar estuary. Ninety two species of commercially important finfishes (12 marine, 63 estuarine cum marine, 9 estuarine and 8 freshwater) were noted in Vellar Estuary by Khan (2015). Fish diversity of Vellar estuary showing decreasing trend as 90 valid species of fishes currently known to occur here. Mahesh and Saravanakumar (2015) identified 45 species belonging to 38 genera, 29 families from Pichavaram mangrove ecosystem. Johnson and Selvaraj (2008) described 23 species of fishes from Manakudy estuary. Kannappan and Karthikeyan (2013) investigated 30 fish species belonging to 13 families and 23 genera in the Manakudy estuary. Pavinkumar *et al.* (2015) recorded 77 number of fish species under 41 families and 11 orders from Manakudy estuary. Present report indicated presence of 91 species of fishes in the Manakudy estuary. Raju *et al.* (2015) described 6 species of fishes from Arasalar estuary.

However diversity of fishes for individual estuaries of Tamil Nadu is lower than Kurup and Samuel (1987); Bijukumar and Sushama (2000); Ghosh *et al.* (2011); Roshith *et al.* (2013). The lower number of species in the estuaries of Tamil Nadu compared to the Cochin estuary, Ponnani estuary, Subarnarekha estuary and Hooghly estuary in India is understandable as the latter are large estuarine systems than the former. Kurup and Samuel (1987) recorded 150 species fishes in Cochin estuary on the west coast of India. Bijukumar and Sushama (2000) indicated 112 species of fishes belonging to 14 orders, 53 families and 80 genera from Ponnani estuary, Kerala. Shrivastava *et al.* (2009) reported 47 fish species belonging to 30 families and 40 genera were recorded from the Krishna estuary. Investigations by Ghosh *et al.* (2011) reported 140 species of fishes belonging to 18 orders and 55 families from Subarnarekha estuary in Orissa. Roshith *et al.* (2013) reported 155 fish species belonging to 49 families and 15 orders from the tidal freshwater zone of the Hooghly estuary.

CONCLUSIONS

Tamil Nadu is endowed with rich estuarine fish diversity. Commercial and recreational fishing could be organized. Available data indicated the lack of scientific studies on estuarine fishes from the different locality of Tamil Nadu. Hence research need to be conducted on fish identification, taxonomic diversity, patterns of distribution and conservation status in Tamil Nadu for planning appropriate conservation strategies.

REFERENCES

Bharadhirajan, P., Murugan, S., Gopalakershanan, A. & Murugesan, P. 2015. Finfish diversity in Coleroon estuary, Southeast coast of India. Indian Journal of Geo-Marine Sciences, 44 (1).

Bijukumar, A. & Sushama, S. 2000. Ichthyofauna of Ponnani estuary, Kerala. The Marine Biological Association of India, 42 (1&2): 182 – 189.

De, D.K. 2011. Estuarine Fisheries. In: 2nd edn. In: Ayyappan, A., Moza, U., Gopalkrishnan, A., Meenakumari, B., Jena, J.K., and Pandey A.K. (eds.), Handbook of Fisheries and Aquaculture, ICAR, New Delhi, 208-237pp.

Ghosh, A. Bhaumik, U. & Satpathy, B.B. 2011. Fish diversity of Subarnarekha estuary in relation to Salinity. J. Inland Fish. Soc. India, 43(1): 51-61.

Jayaram, K.C. 2010. The Freshwater Fishes of the Indian Region - 2nd Edition. Narendra Publishing House, Delhi, 616pp+39pls.

Jha, B.C., Nath, D., Srivastava, N.P. & Satpathy, B.B. 2008. Estuarine Fisheries Management - Options & Strategies. Policy No. 03, Central Inland Fisheriers Rsearch Institute, Barrackpore, Kolkata, 29pp.

Johnson, M. & Selvaraj, D. 2008. Diversity of fishes in Manakudy Estuary, Southwest coast of India. Journal of basic and applied biology. 2(1): 83-86.

Kannappan, T. & Karthikeyan, M.M. 2013. Diversity of fishes in relation to physcio-chemical properties of Manakudy estuary, Southwest coast of India. International Journal of Biodiversity and Conservation, 5(7), pp. 396-407.

Khan, A.S. 2015. Variations in the diversity of commercially important finfishes in Vellar Estuary, south-east coast of India. Indian J. Fish., 62(4): 116-119.

*Kumaran, B., Kambala , S.N. & Nadarajan, J. 2012. Assessment of Ichthyo-faunal Diversity in Giriyampeta Estuary, Yanam (U.T. of Puducherry). Bulletin of Environment, Pharmacology and Life Sciences, 1 (9): 17 – 25.

Kurup, B.M. & Samuel, C.T. 1987. Ecology and fish distribution pattern of a tropical estuary. In: Nair, N.B. (ed), Proc. Natl. Sem. Estuarine, Trivandrum. 1987:339–349.

Mahesh, R. & Saravanakumar, A. 2015. Temporal and spatial variability of fin fish assemblage structure in relation to their environmental parameters in Pichavaram mangrove ecosystem, India. Indian Journal of Geo-Marine Sciences, Vol. 44(6).

Pavinkumar, P. 2014. Fish diversity in selected estuaries of southern Tamil Nadu. Thesis submitted to Tamil Nadu Fisheries Unviversity, Nagapattinam. Pp 122.

Pavinkumar, P. Jawahar, P. & Mogalekar, H.S. 2015. Estuarine fish diversity of Manakudy estuary, Kanyakumari district, Tamil Nadu, India. J. Env. Bio-Sci., 29 (2): 523-528.

Raju, C., Sridharan, G., Mariappan, P. & Chelladurai, G. 2015. Physico-chemical parameters and Ichthyofauna diversity of Arasalar estuary in southeast coast of India. Applied Water Science, DOI 10.1007/s13201-014-0260-0.

Ramaiyan, V. Purushothaman, A. & Nataraian, R. 1987. Check - list of estuarine and marine fishes of Parangipettai (Porto Novo) coastal waters. Matsya, 12-13 1-19, 1986-87.

Ramanujam, M.E. & Anbarasan, R. 2008. A priliminary report on the ichthyofauna of Yedayanthittu Estuary (Tamil Nadu, India) and rivulets draining into it. Journal of Threatened Taxa, 1(5): 287-294.

Ramanujam, M.E., Devi, K.R. & Indra, T.J. 2014. Ichthyofaunal diversity of the Adyar Wetland complex, Chennai, Tamil Nadu, southern India. Journal of Threatened Taxa, 6(4): 5613–5635; http://dx.doi.org/10.11609/JoTT.o2905.5613-35.

Ramesh, R., Nammalwar, P. & Gowri, V.S. 2008. The Database on Coastal Information of Tamilnadu, Environmental Information System (ENVIS) Centre, Department of Environment, Government of Tamilnadu Chennai. Tamil Nadu Forest Department, 153pp.

Remadevi, K., Indra, T. J. & Raghunathan, M.B. 2004. Fishes of Pulicat Lake. Rec. zool. Surv. India: 102 (Part 3-4): 33-42.

Roshith, C.M., Sharma, A.P., Manna, R.K., Satpathy, B.B. and Bhaumik, U. 2013. Ichthyofaunal diversity, assemblage structure and seasonal dynamics in the freshwater tidal stretch of Hooghly estuary along the Gangetic delta. Aquatic Ecosystem Health & Management, 16(4):445–453. http://dx.doi.org/10.1080/14634988.2013.857561

Shrivastava, N. P., Jha, B.C., Nath, D. & Karmakar, H.C. 2009. Krishna estuary - ecology and fisheries. Bull. No. 158, Central Inland Fisheries Research Institute, (Indian Council of Agricultural Research), Barrackpore, Kolkata - 700 120, West Bengal, 43pp.

Sukumaran, M., Muthukumaravel, K. & Kothandapani, S. 2014. Studies on the diversity of Ichthyo - fauna in Muthupet Estuary South east coast of India. International Research in Arts and Sciences, 3, (10), 037-040.

Talwar, P.K. & Jhingran, A.G. 1991. Inland Fishes of India and Adjacent Countries - Volume 1 & 2. Oxford and IBH Publishing Co. Pvt. Ltd., 1158pp.

Talwar, P.K. & Kacker, R.K. 1984. Commercial Sea Fishes of India. Zoological Survey of India, Calcutta, 997pp.

* Referred from predatory journals (Further confirmation is required for the species reported from predatory journals)

Table 1: Fish diversity of selected estuaries of Tamil Nadu with note on distribution, environment, human use or fishery status, maximum size and conservation status.

Order, Family, Scientific name	Distribution of fishes in the Estuaries of Tamil Nadu								Environment	Human use oe fishery status	Maximum size (TL)	IUCN Status	References
	I	II	III	IV	V	VI	VII	VIII					
Class: Chondrichthyes													
Order: Carcharhiniformes													
Family: Carcharinidae													
Scoliodon laticaudus Müller & Henle, 1838	-	-	-	-	-	+	-	-	Marine; Brackish	Commercial; Bait	100.0 cm	NT	Kumaran *et al.* (2012)
Order: Lamniformes													
Family: Lamnidae													
Isurus oxyrinchus Rafinesque, 1810	-	-	-	-	-	+	-	-	Marine	Minor commercial; Gamefish	400 cm	VU	Kumaran *et al.* (2012)
Order: Rajiformes													
Family: Rhinobatidae													
Rhinobatos obtusus Müller & Henle, 1841	+	-	-	-	-	-	-	-	Marine	Fisheries of no interest	93.0 cm	VU	Ramanujam *et al.* (2014)
Class: Osteichthyes													
Order: Albuliformes													
Family: Albulidae													

Order, Family, Scientific name	Distribution of fishes in the Estuaries of Tamil Nadu								Environment	Human use oe fishery status	Maximum size (TL)	IUCN Status	References
	I	II	III	IV	V	VI	VII	VIII					
Albula vulpes (Linnaeus, 1758)	+	-	-	-	-	-	-	-	Marine; Brackish	Minor commercial; Gamefish; Bait	104 cm	NT	Ramanujam *et al.* (2014)
Order: Anguilliforms													
Family: Anguillidae													
Anguilla bengalensis (Gray, 1831)	-	-	-	+	-	+	-	-	Marine; Freshwater; Brackish	Commercial; Aquaculture; Gamefish	200 cm	NT	Kumaran *et al.* (2012); Kannappan & Karthikeyan (2013)
Anguilla bicolor McClelland, 1844	+	-	-	+	+	-	-	-	Marine; Freshwater; Brackish	Minor commercial	123 cm	NT	Ramanujam & Anbarasan (2008); Kannappan & Karthikeyan (2013); Ramanujam *et al.* (2014); Khan (2015); Pavinkumar *et al.* (2015)
Family: Congridae													
Conger cinereus Rüppell, 1830	-	-	+	+	-	-	-	-	Marine; Brackish	Commercial; Gamefish; Ornamental	140 cm	NE	Bharadhirajan *et al.* (2015); Pavinkumar *et al.* (2015)
Family: Muraenesocidae													
Congresox talabon (Cuvier, 1829)	-	-	-	-	+	-	-	-	Marine; Brackish	Minor commercial	80.0 cm	NE	Ramanujam & Anbarasan (2008)
Muraenesox bagio (Hamilton, 1822)	-	+	-	-	+	-	-	-	Marine; Brackish	Commercial; Gamefish	200 cm	NE	Remadevi *et al.* (2004); Ramanujam & Anbarasan (2008)

Order, Family, Scientific name	Distribution of fishes in the Estuaries of Tamil Nadu								Environment	Human use oe fishery status	Maximum size (TL)	IUCN Status	References
	I	II	III	IV	V	VI	VII	VIII					
Family: Ophichthidae													
Bascanichthys deraniyagalai Menon, 1961	-	-	-	-	+	-	-	-	Marine; Brackish	Fisheries of no interest	60.0 cm	NE	Ramanujam & Anbarasan (2008)
Somniosus microcephalus (Bloch & Schneider, 1801)	+	-	-	-	-	-	-	-	Marine; Brackish	Minor commercial; Gamefish	730 cm	NT	Ramanujam *et al.* (2014)
Pisodonophis boro (Hamilton, 1822)	+	-	-	-	-	-	-	-	Marine; Freshwater; Brackish	Minor commercial; Bait	100.0 cm	LC	Ramanujam *et al.* (2014)
Order: Aulopiformes													
Family: Synodontidae													
Harpadon nehereus (Hamilton, 1822)	-	-	-	-	-	+	-	-	Marine; Brackish	Highly commercial	40.0 cm	NE	Kumaran *et al.* (2012)
Saurida tumbil (Bloch, 1795)	-	+	-	-	-	-	-	-	Marine	Commercial	60.0 cm	NE	Remadevi *et al.* (2004);
Synodus sageneus Waite, 1905	-	-	+	-	-	-	-	-	Marine	Commercial	26.0 cm	NE	Mahesh & Saravanakumar (2015)
Order: Atheriniformes													
Family: Atherinidae													
Atherinomorus duodecimalis Valenciennes, 1835	-	+	-	-	-	-	-	-	Marine; Brackish	Bait	11.0 cm SL	NE	Remadevi *et al.* (2004)

Order, Family, Scientific name	Distribution of fishes in the Estuaries of Tamil Nadu								Environment	Human use oe fishery status	Maximum size (TL)	IUCN Status	References
	I	II	III	IV	V	VI	VII	VIII					
Hypoatherina temminckii Bleeker, 1854	-	+	-	-	-	-	-	-	Marine	Subsistence fisheries; Bait	12.0 cm	NE	Remadevi *et al.* (2004)
Order: Beloniformes													
Family: Adrianichthyidae													
Oryzias melastigma (McClelland, 1839)	-	-	-	-	+	-	-	-	Freshwater; Brackish	Ornamental	4.0 cm	NE	Ramanujam & Anbarasan (2008)
Family: Belonidae													
Strongylura strongylura (van Hasselt, 1823)	-	-	-	-	+	-	-	-	Marine; Brackish	Commercial	40.0 cm SL	LC	Ramanujam & Anbarasan (2008)
Strongylura leiura (Bleeker, 1850)	-	-	+	-	-	-	-	-	Marine; Brackish	Commercial; Gamefish	100.0 cm	NE	Mahesh & Saravanakumar (2015)
Tylosurus crocodilus (Péron & Lesueur, 1821)	-	-	-	+	-	-	-	-	Marine	Commercial; Gamefish	150 cm	NE	Pavinkumar *et al.* (2015)
Xenentodon cancila (Hamilton, 1822)	-	+	-	+	+	-	-	-	Marine; Freshwater; Brackish	Minor commercial; Ornamental	40.0 cm	LC	Remadevi *et al.* (2004); Ramanujam & Anbarasan (2008); Pavinkumar *et al.* (2015)
Family: Hemiramphidae													
Hemiramphus far (Forsskål, 1775)	-	-	+	-	-	-	-	-	Marine; Brackish	Commercial; Gamefish; Bait	45.0 cm	NE	Bharadhirajan *et al.* (2015); Khan (2015); Mahesh & Saravanakumar (2015)
Hemiramphus marginatus (Forsskål, 1775)	-	-	-	+	-	-	-	-	Marine	Minor commercial	26.0 cm SL	NE	Pavinkumar *et al.* (2015)

Order, Family, Scientific name	Distribution of fishes in the Estuaries of Tamil Nadu								Environment	Human use oe fishery status	Maximum size (TL)	IUCN Status	References
	I	II	III	IV	V	VI	VII	VIII					
Hyporhamphus limbatus (Valenciennes, 1847)	-	+	+	-	+	-	-	-	Marine; Freshwater; Brackish	Minor commercial	35.0 cm	LC	Remadevi *et al.* (2004); Ramanujam & Anbarasan (2008); Khan (2015)
Order: Clupeiformes													
Family: Chirocentridae													
Chirocentrus dorab (Forsskål, 1775)	-	-	-	-	-	+	-	-	Marine; Brackish	Commercial; Gamefish; Bait	100.0 cm SL	NE	Kumaran *et al.* (2012)
Chirocentrus nudus Swainson, 1839	-	-	-	-	-	+	-	-	Marine	Highly commercial	100.0 cm SL	NE	Kumaran *et al.* (2012)
Family: Clupeidae													
Anodontostoma chacunda (Hamilton, 1822)	-	+	+	-	+	-	-	-	Marine; Freshwater; Brackish	Commercial	22.0 cm SL	NE	Remadevi *et al.* (2004); Ramanujam & Anbarasan (2008); Khan (2015); Mahesh & Saravanakumar (2015)
Escualosa thoracata (Valenciennes, 1847)	+	-	+	-	-	-	-	-	Marine; Freshwater; Brackish	Commercial	10.0 cm SL	NE	Ramanujam *et al.* (2014); Bharadhirajan *et al.* (2015); Khan (2015)
Herklotsichthys quadrimaculatus (Rüppell, 1837)	-	+	-	-	-	-	-	-	Marine; Freshwater; Brackish	Minor commercial; Bait	25.0 cm SL	NE	Remadevi *et al.* (2004)

Order, Family, Scientific name	Distribution of fishes in the Estuaries of Tamil Nadu								Environment	Human use oe fishery status	Maximum size (TL)	IUCN Status	References
	I	II	III	IV	V	VI	VII	VIII					
Hilsa kelee (Cuvier, 1829)	+	+	+	+	-	-	-	-	Marine; Freshwater; Brackish	Highly commercial; Bait	35.0 cm	NE	Remadevi *et al.* (2004); Kannappan & Karthikeyan (2013); Ramanujam *et al.* (2014); Bharadhirajan *et al.* (2015); Khan (2015)
Nematalosa nasus (Bloch, 1795)	+	+	+	+	+	-	-	-	Marine; Freshwater; Brackish	Minor commercial	22.0 cm	LC	Remadevi *et al.* (2004); Ramanujam & Anbarasan (2008); Ramanujam *et al.* (2014); Bharadhirajan *et al.* (2015); Khan (2015); Pavinkumar *et al.* (2015)
Sardinella albella (Valenciennes, 1847)	-	-	+	+	-	+	-	-	Marine	Commercial	14.0 cm SL	LC	Kumaran *et al.* (2012); Mahesh & Saravanakumar (2015); Pavinkumar *et al.* (2015)
Sardinella fimbriata (Valenciennes, 1847)	-	-	+	-	-	+	-	-	Marine; Brackish	Commercial	13.0 cm SL	NE	Kumaran *et al.* (2012); Bharadhirajan *et al.* (2015); Mahesh & Saravanakumar (2015); Khan (2015);
Sardinella gibbosa (Bleeker, 1849)	+	-	-	-	-	+	-	-	Marine	Highly commercial	17.0 cm SL	NE	Kumaran *et al.* (2012); Ramanujam *et al.* (2014)
Sardinella longiceps Valenciennes, 1847	-	-	+	-	-	+	-	-	Marine	Highly commercial	23.0 cm SL	LC	Kumaran *et al.* (2012); Bharadhirajan *et al.* (2015); Khan (2015)
Sardinella sindensis (Day, 1878)	-	-	+	-	-	-	-	-	Marine	Minor commercial	17.0 cm SL	NE	Khan (2015)

Order, Family, Scientific name	Distribution of fishes in the Estuaries of Tamil Nadu								Environment	Human use oe fishery status	Maximum size (TL)	IUCN Status	References
	I	II	III	IV	V	VI	VII	VIII					
Tenualosa ilisha (Hamilton, 1822)	-	-	-	+	-	+	-	-	Marine; Freshwater; Brackish	Minor commercial; Aquaculture	60.0 cm SL	LC	Kumaran *et al.* (2012); Pavinkumar *et al.* (2015)
Tenualosa toli (Valenciennes, 1847)	-	-	-	-	-	+	-	-	Marine; Freshwater; Brackish	Highly commercial	60.0 cm	NE	Kumaran *et al.* (2012)
Family: Dussumieriidae													
Dussumieria acuta Valenciennes, 1847	-	-	-	+	-	-	-	-	Marine; Freshwater; Brackish	Highly commercial	20.0 cm SL	NE	Pavinkumar *et al.* (2015)
Family: Engraulidae													
Coilia dussumieri Valenciennes, 1848	-	-	+	-	+	+	-	-	Marine; Freshwater; Brackish	Commercial	20.0 cm SL	NE	Ramanujam & Anbarasan (2008); Kumaran *et al.* (2012); Bharadhirajan *et al.* (2015); Khan (2015)
Setipinna taty (Valenciennes, 1848)	-	-	+	-	-	-	-	-	Marine; Brackish	Minor commercial	15.3 cm	NE	Khan (2015)
Stolephorus andhraensis Babu Rao, 1966	-	+	-	-	-	-	-	-	Marine	Bait	5.0 cm SL	NE	Remadevi *et al.* (2004)
Stolephorus commersonnii Lacepède, 1803	-	-	+	+	-	+	+	-	Marine; Brackish	Commercial	10.0 cm SL	NE	Kumaran *et al.* (2012); Sukumaran *et al.* (2014); Bharadhirajan *et al.* (2015); Khan (2015); Pavinkumar *et al.* (2015)

Order, Family, Scientific name	Distribution of fishes in the Estuaries of Tamil Nadu								Environment	Human use oe fishery status	Maximum size (TL)	IUCN Status	References
	I	II	III	IV	V	VI	VII	VIII					
Stolephorus indicus (van Hasselt, 1823)	-	-	+	+	+	-	-	-	Marine; Brackish	Minor commercial; Bait	15.5 cm SL	NE	Ramanujam & Anbarasan (2008); Mahesh & Saravanakumar (2015); Pavinkumar *et al.* (2015)
Stolephorus tri (Bleeker, 1852)	-	-	-	-	+	-	-	-	Marine; Brackish	Subsistence fisheries	9.5 cm SL	NE	Ramanujam & Anbarasan (2008)
Thryssa dussumieri (Valenciennes, 1848)	-	-	+	-	-	-	-	-	Marine; Brackish	Commercial	11.0 cm SL	NE	Bharadhirajan *et al.* (2015)
Thryssa hamiltonii Gray, 1835	-	+	-	-	-	-	-	-	Marine; Brackish	Commercial	27.0 cm SL	NE	Remadevi *et al.* (2004)
Thryssa kammalensis (Bleeker, 1849)	-	+	-	-	-	-	-	-	Marine; Brackish	Subsistence fisheries	15.0 cm SL	NE	Remadevi *et al.* (2004)
Thryssa malabarica (Bloch, 1795)	+	+	+	-	-	-	-	-	Marine; Brackish	Commercial	17.5 cm SL	NE	Remadevi *et al.* (2004); Ramanujam *et al.* (2014); Bharadhirajan *et al.* (2015); Khan (2015)
Thryssa mystax (Bloch & Schneider, 1801)	+	+	+	-	+	-	-	-	Marine; Brackish	Commercial	15.5 cm SL	LC	Remadevi *et al.* (2004); Ramanujam & Anbarasan (2008); Ramanujam *et al.* (2014); Bharadhirajan *et al.* (2015); Mahesh & Saravanakumar (2015); Khan (2015)
Thryssa purava (Hamilton, 1822)	-	+	-	-	+	-	-	-	Marine; Brackish	Minor commercial	15.5 cm SL	NE	Remadevi *et al.* (2004); Ramanujam & Anbarasan (2008); Ramanujam *et al.* (2014)

Order, Family, Scientific name	Distribution of fishes in the Estuaries of Tamil Nadu								Environment	Human use oe fishery status	Maximum size (TL)	IUCN Status	References
	I	II	III	IV	V	VI	VII	VIII					
Thryssa vitrirostris (Gilchrist & Thompson, 1908)	-	-	+	+	-	-	-	-	Marine; Brackish	Minor commercial; Bait	20.0 cm	NE	Bharadhirajan *et al.* (2015); Khan (2015); Pavinkumar *et al.* (2015)
Family: Pristigasteridae													
Ilisha kampeni (Weber & de Beaufort, 1913)	-	-	+	-	-	-	-	-	Marine; Freshwater; Brackish	Minor commercial	15.0 cm SL	NE	Bharadhirajan *et al.* (2015); Khan (2015)
Ilisha melastoma (Bloch & Schneider, 1801)	-	-	+	-	-	-	-	-	Marine; Brackish	Minor commercial	17.0 cm SL	NE	Khan (2015)
Opisthopterus tardoore (Cuvier, 1829)	-	-	+	-	-	-	-	-	Marine; Brackish	Minor commercial	20.0 cm SL	NE	Khan (2015)
Order: Cyprinidontiformes													
Family: Aplocheilidae													
Aplocheilus parvus (Sundara Raj, 1916)	-	-	-	-	+	-	-	-	Freshwater; Brackish	Ornamental	6.3 cm	NE	Ramanujam & Anbarasan (2008)
Order: Cypriniformes													
Family: Cobitidae													
Lepidocephalichthys thermalis (Valenciennes, 1846)	-	-	-	-	+	-	-	-	Freshwater	Ornamental	38.0 cm SL	LC	Ramanujam & Anbarasan (2008)
Family: Cyprinidae													

Order, Family, Scientific name	Distribution of fishes in the Estuaries of Tamil Nadu								Environment	Human use oe fishery status	Maximum size (TL)	IUCN Status	References
	I	II	III	IV	V	VI	VII	VIII					
Amblypharyngodon microlepis (Bleeker, 1853)	-	-	-	-	+	-	-	-	Freshwater	Ornamental	10.0 cm	LC	Ramanujam & Anbarasan (2008)
Catla catla (Hamilton, 1822)	+	-	-	-	-	-	-	-	Freshwater; Brackish	Highly commercial; Aquaculture; Gamefish	182 cm	LC	Ramanujam *et al.* (2014)
Cirrhinus mrigala (Hamilton, 1822)	-	-	+	-	-	+	-	-	Freshwater	Commercial	99.0 cm	LC	Kumaran *et al.* (2012); Khan (2015)
Cirrhinus reba (Hamilton, 1822)	-	-	-	-	-	+	-	-	Freshwater	Commercial	30.0 cm	NE	Kumaran *et al.* (2012);
Cyprinus carpio Linnaeus, 1758	-	-	+	-	-	-	-	-	Freshwater; Brackish	Highly commercial; Aquaculture; Gamefish; Ornamental	120 cm	VU	Khan (2015)
Esomus danricus (Hamilton, 1822)	+	-	-	-	+	-	-	-	Freshwater; Brackish	Minor commercial; Ornamental	13.0 cm	LC	Ramanujam & Anbarasan (2008); Ramanujam *et al.* (2014)
Bangana dero (Hamilton, 1822)	+	-	-	-	-	-	-	-	Freshwater	Commercial; Bait	75.0 cm	LC	Ramanujam *et al.* (2014)
Labeo bata (Hamilton, 1822)	-	-	-	-	-	+	-	-	Freshwater	Commercial; Aquaculture	61.0 cm	LC	Kumaran *et al.* (2012)
Labeo calbasu (Hamilton, 1822)	-	-	-	-	-	+	-	-	Freshwater; Brackish	Commercial; Aquaculture	90.0 cm	LC	Kumaran *et al.* (2012)

Order, Family, Scientific name	Distribution of fishes in the Estuaries of Tamil Nadu								Environment	Human use oe fishery status	Maximum size (TL)	IUCN Status	References
	I	II	III	IV	V	VI	VII	VIII					
Labeo rohita (Hamilton, 1822)	-	-	+	-	-	+	-	-	Freshwater; Brackish	Highly commercial; Aquaculture; Gamefish	200 cm	LC	Kumaran *et al.* (2012); Khan (2015)
Puntius dorsalis (Jerdon, 1849)	-	-	-	-	+	-	-	-	Freshwater	Minor commercial; Ornamental	25.0 cm	LC	Ramanujam & Anbarasan (2008)
Puntius sophore (Hamilton, 1822)	-	-	-	-	+	-	-	-	Freshwater; Brackish	Ornamental	20.0 cm	LC	Ramanujam & Anbarasan (2008)
Puntius vittatus Day, 1865	-	-	-	-	+	-	-	-	Freshwater; Brackish	Ornamental; Bait	5.0 cm	LC	Ramanujam & Anbarasan (2008)
Salmophasia balookee (Sykes, 1839)	-	-	-	-	+	-	-	-	Freshwater	Minor commercial	15.0 cm	LC	Ramanujam & Anbarasan (2008)
Rasbora daniconius (Hamilton, 1822)	-	-	-	-	+	-	-	-	Freshwater; Brackish	Minor commercial; Ornamental	15.0 cm	LC	Ramanujam & Anbarasan (2008)
Order: Elopiformes													
Family: Elopidae													
Elops machnata (Forsskål, 1775)	+	-	+	+	-	-	-	-	Marine; Brackish	Commercial; Gamefish	118 cm FL	LC	Kannappan & Karthikeyan (2013); Ramanujam *et al.* (2014); Bharadhirajan *et al.* (2015); Mahesh & Saravanakumar (2015); Khan (2015); Pavinkumar *et al.* (2015)

Order, Family, Scientific name	Distribution of fishes in the Estuaries of Tamil Nadu								Environment	Human use oe fishery status	Maximum size (TL)	IUCN Status	References
	I	II	III	IV	V	VI	VII	VIII					
Family: Megalopidae													
Megalops cyprinoides (Broussonet, 1782)	+	-	+	+	+	-	-	-	Marine; Freshwater; Brackish	Minor commercial; Aquaculture; Gamefish	150 cm	DD	Ramanujam & Anbarasan (2008); Ramanujam *et al.* (2014); Bharadhirajan *et al.* (2015); Khan (2015); Pavinkumar *et al.* (2015)
Order: Gonorhynchiformes													
Family: Chanidae													
Chanos chanos (Forsskål, 1775)	+	-	+	+	+	-	+	+	Marine; Freshwater; Brackish	Highly commercial; Aquaculture; Gamefish; Bait	180 cm SL	NE	Johnson & Selvaraj (2008); Ramanujam & Anbarasan (2008); Kannappan & Karthikeyan (2013); Ramanujam *et al.* (2014); Sukumaran *et al.* (2014); Bharadhirajan *et al.* (2015); Khan (2015); Mahesh & Saravanakumar (2015); Pavinkumar *et al.* (2015); Raju *et al.* (2015)
Order: Mugiliformes													
Family: Mugilidae													
Chelon macrolepis (Smith, 1846)	+	-	+	-	-	-	-	-	Marine; Freshwater; Brackish	Commercial; Aquaculture; Gamefish	60.0 cm SL	LC	Ramanujam *et al.* (2014); Bharadhirajan *et al.* (2015); Khan (2015); Mahesh & Saravanakumar (2015)

Order, Family, Scientific name	Distribution of fishes in the Estuaries of Tamil Nadu								Environment	Human use oe fishery status	Maximum size (TL)	IUCN Status	References
	I	II	III	IV	V	VI	VII	VIII					
Chelon parsia (Hamilton, 1822)	+	+	+	+	+	-	+	+	Marine; Freshwater; Brackish	Commercial; Aquaculture	16.0 cm	NE	Remadevi *et al.* (2004); Ramanujam & Anbarasan (2008); Kannappan & Karthikeyan (2013); Ramanujam *et al.* (2014); Sukumaran *et al.* (2014); Bharadhirajan *et al.* (2015); Khan (2015); Mahesh & Saravanakumar (2015); Pavinkumar *et al.* (2015); Raju *et al.* (2015)
Chelon planiceps (Valenciennes, 1836)	+	+	+	+	-	-	-	-	Marine; Freshwater; Brackish	Commercial; Aquaculture	70.0 cm	DD	Remadevi *et al.* (2004); Kannappan & Karthikeyan (2013); Ramanujam *et al.* (2014); Bharadhirajan *et al.* (2015); Khan (2015); Pavinkumar *et al.* (2015)
Chelon subviridis (Valenciennes, 1836)	-	+	+	+	-	-	-	-	Marine; Freshwater; Brackish	Commercial; Aquaculture; Bait	40.0 cm SL	NE	Remadevi *et al.* (2004); Bharadhirajan *et al.* (2015); Khan (2015); Pavinkumar *et al.* (2015)
Ellochelon vaigiensis (Quoy & Gaimard, 1825)	-	+	-	-	-	-	-	-	Marine; Freshwater; Brackish	Commercial; Aquaculture; Ornamental; Bait	63.0 cm	LC	Remadevi *et al.* (2004);

Order, Family, Scientific name	Distribution of fishes in the Estuaries of Tamil Nadu								Environment	Human use oe fishery status	Maximum size (TL)	IUCN Status	References
	I	II	III	IV	V	VI	VII	VIII					
Mugil cephalus Linnaeus, 1758	+	+	+	+	+	+	+	+	Marine; Freshwater; Brackish	Highly commercial; Aquaculture; Gamefish; Bait	100.0 cm SL	LC	Remadevi *et al.* (2004); Johnson & Selvaraj (2008); Ramanujam & Anbarasan (2008); Kumaran *et al.* (2012); Ramanujam *et al.* (2014); Sukumaran *et al.* (2014); Khan (2015); Mahesh & Saravanakumar (2015); Pavinkumar *et al.* (2015); Raju *et al.* (2015)
Moolgarda seheli (Forsskål, 1775)	-	+	+	+	-	-	-	-	Marine; Freshwater; Brackish	Commercial	60.0 cm	NE	Remadevi *et al.* (2004); Ramanujam & Anbarasan (2008); Bharadhirajan *et al.* (2015); Khan (2015); Pavinkumar *et al.* (2015)
Order: Perciformes													
Family: Acanthuridae													
Acanthurus mata (Cuvier, 1829)	-	-	+	+	-	-	-	-	Marine	Commercial; Ornamental	50.0 cm	LC	Khan (2015); Pavinkumar *et al.* (2015)
Ctenochaetus striatus (Quoy & Gaimard, 1825)	-	-	+	-	-	-	-		Marine	Commercial; Ornamental	26.0 cm	LC	Mahesh & Saravanakumar (2015)
Family: Ambassidae													

Order, Family, Scientific name	Distribution of fishes in the Estuaries of Tamil Nadu								Environment	Human use oe fishery status	Maximum size (TL)	IUCN Status	References
	I	II	III	IV	V	VI	VII	VIII					
Ambassis ambassis (Lacepède, 1802)	+	-	+	-	+	-	+	-	Marine; Freshwater; Brackish	Ornamental	15.0 cm	LC	Ramanujam & Anbarasan (2008); Ramanujam *et al.* (2014); Sukumaran *et al.* (2014); Bharadhirajan *et al.* (2015); Mahesh & Saravanakumar (2015); Khan (2015)
Ambassis gymnocephalus (Lacepède, 1802)	-	-	+	-	-	-	-	-	Marine; Freshwater; Brackish	Minor commercial; Bait	16.0 cm	LC	Khan (2015)
Chanda nama Hamilton, 1822	+	-	-	+	-	-	-	-	Freshwater; Brackish	Minor commercial; Ornamental	11.0 cm	LC	Ramanujam *et al.* (2014); Pavinkumar *et al.* (2015)
Parambassis ranga (Hamilton, 1822)	+	-	-	-	+	-	-	-	Freshwater; Brackish	Subsistence fisheries; Ornamental	8.0 cm	LC	Ramanujam & Anbarasan (2008); Ramanujam *et al.* (2014)
Family: Anabantidae													
Anabas testudineus (Bloch, 1792)	+	-	+	+	+	-	-	-	Freshwater; Brackish	Commercial; Aquaculture; Ornamental	25.0 cm	DD	Ramanujam & Anbarasan (2008); Kannappan & Karthikeyan (2013); Ramanujam *et al.* (2014); Khan (2015)
Family: Apogonidae													
Apogonichthyoides niger (Döderlein, 1883)	-	-	+	-	-	-	-	-	Marine	Fisheries of no interest	8.0 cm	NE	Khan (2015)
Family: Carangidae													

Order, Family, Scientific name	Distribution of fishes in the Estuaries of Tamil Nadu								Environment	Human use oe fishery status	Maximum size (TL)	IUCN Status	References
	I	II	III	IV	V	VI	VII	VIII					
Alectis ciliaris (Bloch, 1787)	-	-	+	-	-	-	-	-	Marine	Minor commercial; Gamefish; Ornamental	150 cm	LC	Bharadhirajan *et al.* (2015); Khan (2015)
Alectis indica (Rüppell, 1830)	-	-	+	-	-	-	-	-	Marine; Brackish	Commercial; Gamefish; Ornamental	165 cm	NE	Bharadhirajan *et al.* (2015); Khan (2015)
Alepes djedaba (Forsskål, 1775)	-	-	+	-	-	-	-	-	Marine	Commercial; Gamefish	40.0 cm	NE	Mahesh & Saravanakumar (2015)
Alepes kleinii (Bloch, 1793)	+	-	+	-	-	-	-	-	Marine	Minor commercial	16.0 cm FL	NE	Ramanujam *et al.* (2014); Khan (2015)
Atule mate (Cuvier, 1833)	-	-	-	-	+	-	-	-	Marine; Brackish	Minor commercial; Gamefish	30.0 cm	NE	Ramanujam & Anbarasan (2008)
Carangoides chrysophrys (Cuvier, 1833)	-	-	-	-	+	-	-	-	Marine; Brackish	Commercial; Gamefish	72.0 cm FL	NE	Ramanujam & Anbarasan (2008)
Carangoides coeruleopinnatus (Rüppell, 1830)	-	-	-	+	-	-	-	-	Marine	Minor commercial; Gamefish	41.0 cm	NE	Pavinkumar *et al.* (2015)
Carangoides ferdau (Forsskål, 1775)	-	-	-	+	-	-	-	-	Marine; Brackish	Commercial; Gamefish	70.0 cm	NE	Pavinkumar *et al.* (2015)
Carangoides malabaricus (Bloch & Schneider, 1801)	-	-	+	-	+	-	-	-	Marine	Commercial; Gamefish	60.0 cm	NE	Ramanujam & Anbarasan (2008); Bharadhirajan *et al.* (2015); Khan (2015)

Order, Family, Scientific name	Distribution of fishes in the Estuaries of Tamil Nadu								Environment	Human use oe fishery status	Maximum size (TL)	IUCN Status	References
	I	II	III	IV	V	VI	VII	VIII					
Caranx heberi (Bennett, 1830)	-	-	+	-	-	-	-	-	Marine; Brackish	Minor commercial; Gamefish	88.0 cm	NE	Khan (2015)
Caranx ignobilis (Forsskål, 1775)	+	-	+	+	-	+	-	-	Marine; Brackish	Commercial; Aquaculture; Gamefish; Ornamental	170 cm	NE	Kumaran *et al.* (2012); Ramanujam *et al.* (2014); Khan (2015); Pavinkumar *et al.* (2015)
Caranx sexfasciatus Quoy & Gaimard, 1825	+	-	+	+	-	-	-	-	Marine; Freshwater; Brackish	Commercial; Gamefish	120 cm	LC	Ramanujam *et al.* (2014); Bharadhirajan *et al.* (2015); Khan (2015); Pavinkumar *et al.* (2015)
Caranx tille Cuvier, 1833	+	-	-	-	-	-	-	-	Marine	Minor commercial; Gamefish	80.0 cm	NE	Ramanujam *et al.* (2014)
Parastromateus niger (Bloch, 1795)	-	-	+	-	-	+	-	-	Marine; Brackish	Highly commercial	75.0 cm	NE	Kumaran *et al.* (2012); Khan (2015)
Scomberoides commersonnianus Lacepède, 1801	-	-	-	+	-	-	-	-	Marine; Brackish	Minor commercial; Gamefish	120 cm	NE	Pavinkumar *et al.* (2015)
Scomberoides tol (Cuvier, 1832)	-	-	+	+	-	-	-	-	Marine; Brackish	Minor commercial; Gamefish	60.0 cm	NE	Bharadhirajan *et al.* (2015); Khan (2015); Pavinkumar *et al.* (2015)
Trachinotus africanus Smith, 1967	-	-	-	+	-	-	-	-	Marine; Brackish	Minor commercial; Gamefish; Ornamental	92.0 cm	NE	Pavinkumar *et al.* (2015)

Order, Family, Scientific name	Distribution of fishes in the Estuaries of Tamil Nadu								Environment	Human use oe fishery status	Maximum size (TL)	IUCN Status	References
	I	II	III	IV	V	VI	VII	VIII					
Uraspis helvola (Forster, 1801)	-	-	+	-	-	-	-	-	Marine	Minor commercial; Gamefish	58.0 cm	NE	Mahesh & Saravanakumar (2015)
Family: Cepolidae													
Acanthocepola abbreviata (Valenciennes, 1835)	-	-	+	-	-	-	-	-	Marine	Bait	30.0 cm	NE	Khan (2015)
Family: Channidae													
Channa orientalis Bloch & Schneider, 1801	-	-	-	+	-	-	-	-	Freshwater; Brackish	Minor commercial; Ornamental	33.0 cm	NE	Kannappan & Karthikeyan (2013);
Channa punctata (Bloch, 1793)	-	-	-	+	+	-	-	-	Freshwater; Brackish	Commercial; Aquaculture; Ornamental; Bait	31.0 cm	LC	Ramanujam & Anbarasan (2008); Kannappan & Karthikeyan (2013); Pavinkumar *et al.* (2015)
Channa striata (Bloch, 1793)	-	-	-	-	+	-	-	-	Freshwater; Brackish	Highly commercial; Aquaculture; Ornamental	100.0 cm SL	LC	Ramanujam & Anbarasan (2008)
Family: Cichilidae													

Order, Family, Scientific name	Distribution of fishes in the Estuaries of Tamil Nadu								Environment	Human use oe fishery status	Maximum size (TL)	IUCN Status	References
	I	II	III	IV	V	VI	VII	VIII					
Etroplus maculatus (Bloch, 1795)	+	-	+	+	+	-	+	-	Freshwater; Brackish	Commercial; Ornamental	8.0 cm	LC	Johnson & Selvaraj (2008); Ramanujam & Anbarasan (2008); Kannappan & Karthikeyan (2013); Ramanujam *et al.* (2014); Sukumaran *et al.* (2014); Bharadhirajan *et al.* (2015); Pavinkumar *et al.* (2015)
Etroplus suratensis (Bloch, 1790)	-	+	+	+	+	-	+	-	Freshwater; Brackish	Commercial; Aquaculture; Ornamental	40.0 cm	LC	Remadevi *et al.* (2004); Johnson & Selvaraj (2008); Ramanujam & Anbarasan (2008); Kannappan & Karthikeyan (2013); Sukumaran *et al.* (2014); Khan (2015); Mahesh & Saravanakumar (2015); Pavinkumar *et al.* (2015)
Oreochromis mossambicus (Peters, 1852)	+	+	+	+	+	-	+	+	Freshwater; Brackish	Highly commercial; Aquaculture; Gamefish; Ornamental	39.0 cm SL	NT	Remadevi *et al.* (2004); Johnson & Selvaraj (2008); Ramanujam & Anbarasan (2008); Kannappan & Karthikeyan (2013); Ramanujam *et al.* (2014); Sukumaran *et al.* (2014); Mahesh & Saravanakumar (2015); Khan (2015); Pavinkumar *et al.* (2015); Raju *et al.* (2015)

Order, Family, Scientific name	Distribution of fishes in the Estuaries of Tamil Nadu								Environment	Human use oe fishery status	Maximum size (TL)	IUCN Status	References
	I	II	III	IV	V	VI	VII	VIII					
Oreochromis niloticus (Linnaeus, 1758)	-	-	-	+	-	-	-	-	Freshwater; Brackish	Highly commercial; Aquaculture	60.0 cm SL	NE	Kannappan & Karthikeyan (2013);
Family: Drepanidae													
Drepane punctata (Linnaeus, 1758)	-	+	+	-	-	-	-	-	Marine; Brackish	Commercial; Ornamental	50.0 cm	NE	Remadevi *et al.* (2004); Bharadhirajan *et al.* (2015); Khan (2015)
Family: Gerreidae													
Gerres erythrourus (Bloch, 1791)	-	+	+	+	+	-	-	-	Marine; Brackish	Minor commercial	30.0 cm	NE	Remadevi *et al.* (2004); Ramanujam & Anbarasan (2008); Kannappan & Karthikeyan (2013); Bharadhirajan *et al.* (2015); Khan (2015); Pavinkumar *et al.* (2015)
Gerres filamentosus Cuvier, 1829	+	+	+	+	+	-	+	-	Marine; Freshwater; Brackish	Minor commercial	35.0 cm	LC	Remadevi *et al.* (2004); Johnson & Selvaraj (2008); Ramanujam & Anbarasan (2008); Ramanujam *et al.* (2014); Sukumaran *et al.* (2014); Bharadhirajan *et al.* (2015); Mahesh & Saravanakumar (2015); Khan (2015); Pavinkumar *et al.* (2015)
Gerres limbatus Cuvier, 1830	-	+	+	-	-	-	-	-	Marine; Brackish	Commercial	15.0 cm	LC	Remadevi *et al.* (2004); Khan (2015)

Order, Family, Scientific name	Distribution of fishes in the Estuaries of Tamil Nadu								Environment	Human use oe fishery status	Maximum size (TL)	IUCN Status	References
	I	II	III	IV	V	VI	VII	VIII					
Gerres longirostris (Lacepède, 1801)	+	+	-	-	+	-	-	-	Marine; Freshwater; Brackish	Commercial	44.5 cm	NE	Remadevi *et al.* (2004); Ramanujam & Anbarasan (2008); Ramanujam *et al.* (2014)
Gerres methueni Regan, 1920	-	-	+	-	-	-	-	-	Marine; Brackish	Commercial	30.0 cm	NE	Khan (2015)
Gerres oyena (Forsskål, 1775)	+	+	-	+	-	-	-	-	Marine; Brackish	Commercial	30.0 cm	NE	Remadevi *et al.* (2004); Ramanujam *et al.* (2014); Pavinkumar *et al.* (2015)
Gerres setifer (*Hamilton, 1822)*	-	+	-	+		-	-	-	Marine; Brackish	Commercial	15.0 cm SL	NE	Remadevi *et al.* (2004); Pavinkumar *et al.* (2015)
Family: Gobiidae													
Acentrogobius cyanomos (Bleeker, 1849)	-	+	-	-	-	-	-	-	Marine; Brackish	Fisheries of no interest	11.5 cm	NE	Remadevi *et al.* (2004);
Acentrogobius ennorensis Menon & Rema Devi, 1980	-	-	-	+	-	-	-	-	Marine; Brackish	Fisheries of no interest	5.8 cm SL	NE	Pavinkumar *et al.* (2015)
Acentrogobius viridipunctatus (Valenciennes, 1837)	-	+	-	-	-	-	-	-	Marine; Freshwater; Brackish	Fisheries of no interest	16.5 cm	NE	Remadevi *et al.* (2004);
Arcygobius baliurus (Valenciennes, 1837)	-	-	+	-	-	-	-	-	Marine; Brackish	Fisheries of no interest	10.2 cm SL	NE	Mahesh & Saravanakumar (2015)
Boleophthalmus boddarti (Pallas, 1770)	-	+	-	-	+	-	-	-	Marine; Freshwater; Brackish	Fisheries of no interest	22.0 cm	LC	Remadevi *et al.* (2004); Ramanujam & Anbarasan (2008)

Order, Family, Scientific name	Distribution of fishes in the Estuaries of Tamil Nadu								Environment	Human use oe fishery status	Maximum size (TL)	IUCN Status	References
	I	II	III	IV	V	VI	VII	VIII					
Favonigobius reichei (Bleeker, 1854)	-	+	-	-	-	-	-	-	Marine; Freshwater; Brackish	Fisheries of no interest	8.3 cm	LR/ lc	Remadevi *et al.* (2004);
Glossogobius giuris (Hamilton, 1822)	+	+	-	-	+	-	-	-	Marine; Freshwater; Brackish	Minor commercial; Aquaculture; Ornamental	50.0 cm SL	LC	Remadevi *et al.* (2004); Ramanujam & Anbarasan (2008); Ramanujam *et al.* (2014)
Oxyurichthys microlepis (Bleeker, 1849)	-	+	+	-	-	-	-	-	Marine; Brackish	Minor commercial; Ornamental	13.5 cm	NE	Remadevi *et al.* (2004); Mahesh & Saravanakumar (2015)
Oxyurichthys ophthalmonema (Bleeker, 1856)	-	-	+	-	-	-	-	-	Marine; Freshwater; Brackish	Fisheries of no interest	18.0 cm	NE	Mahesh & Saravanakumar (2015)
Parachaeturichthys polynema (Bleeker, 1853)	-	+	-	-	-	-	-	-	Marine	Fisheries of no interest	15.0 cm	NE	Remadevi *et al.* (2004);
Paratrypauchen microcephalus (Bleeker, 1860)	-	+	-	-	-	-	-	-	Marine; Brackish	Minor commercial; Ornamental	18.0 cm	NE	Remadevi *et al.* (2004);
Periophthalmus variabilis Eggert, 1935	-	-	-	-	+	-	-	-	Marine; Brackish	Fisheries of no interest	6.4 cm SL	NE	Ramanujam & Anbarasan (2008)
Psammogobius biocellatus (Valenciennes, 1837)	-	-	+	-	-	-	-	-	Marine; Freshwater; Brackish	Fisheries of no interest	12.0 cm	LC	Remadevi *et al.* (2004); Mahesh & Saravanakumar (2015)

Order, Family, Scientific name	Distribution of fishes in the Estuaries of Tamil Nadu								Environment	Human use oe fishery status	Maximum size (TL)	IUCN Status	References
	I	II	III	IV	V	VI	VII	VIII					
Yongeichthys criniger (Valenciennes, 1837)	-	+	-	-	+	-	-	-	Marine; Brackish	Fisheries of no interest	15.0 cm SL	NE	Remadevi *et al.* (2004); Ramanujam & Anbarasan (2008)
Family: Haemulidae													
Plectorhinchus gibbosus (Lacepède, 1802)	-	-	-	+	-	-	-	-	Marine; Freshwater; Brackish	Commercial; Gamefish	75.0 cm	LC	Pavinkumar *et al.* (2015)
Pomadasys argenteus (Forsskål, 1775)	-	+	-	-	-	-	-	-	Marine; Freshwater; Brackish	Commercial	70.0 cm	LC	Remadevi *et al.* (2004)
Pomadasys kaakan (Cuvier, 1830)	-	+	+	-	-	-	-	-	Marine; Brackish	Commercial; Aquaculture; Gamefish	80.0 cm	NE	Remadevi *et al.* (2004); Bharadhirajan *et al.* (2015); Khan (2015); Mahesh & Saravanakumar (2015)
Pomadasys maculatus (Bloch, 1793)	-	-	+	-	-	-	-	-	Marine; Brackish	Commercial	59.3 cm	LC	Bharadhirajan *et al.* (2015); Khan (2015)
Family: Lactariidae													
Lactarius lactarius (Bloch & Schneider, 1801)	-	+	-	+	-	-	-	-	Marine; Brackish	Commercial	40.0 cm	NE	Remadevi *et al.* (2004); Johnson & Selvaraj (2008)
Family: Leiognathidae													

Order, Family, Scientific name	Distribution of fishes in the Estuaries of Tamil Nadu								Environment	Human use oe fishery status	Maximum size (TL)	IUCN Status	References
	I	II	III	IV	V	VI	VII	VIII					
Eubleekeria splendens (Cuvier, 1829)	+	+	+	+	+	-	-	-	Marine; Brackish	Commercial	17.0 cm	LC	Remadevi *et al.* (2004); Ramanujam & Anbarasan (2008); Ramanujam *et al.* (2014); Bharadhirajan *et al.* (2015); Khan (2015); Pavinkumar *et al.* (2015)
Gazza minuta (Bloch, 1795)	+	+	-	-	+	-	-	-	Marine; Brackish	Commercial	21.0 cm FL	LC	Remadevi *et al.* (2004); Ramanujam & Anbarasan (2008); Ramanujam *et al.* (2014)
Karalla daura (Cuvier, 1829)	-	-	-	+	-	-	-	-	Marine	Commercial	14.0 cm	NE	Remadevi *et al.* (2004); Pavinkumar *et al.* (2015)
Karalla dussumieri (Valenciennes, 1835)	-	+	-	-	-	-	-	-	Marine; Brackish	Commercial	14.0 cm	NE	Remadevi *et al.* (2004);
Leiognathus brevirostris (Valenciennes, 1835)	-	+	-	+	+	-	-	-	Marine; Brackish	Commercial	13.5 cm	NE	Remadevi *et al.* (2004); Ramanujam & Anbarasan (2008); Pavinkumar *et al.* (2015)
Leiognathus equulus (Forsskål, 1775)	-	+	+	+	-	-	+	-	Marine; Freshwater; Brackish	Minor commercial; Aquaculture	28.0 cm	LC	Remadevi *et al.* (2004); Johnson & Selvaraj (2008); Sukumaran *et al.* (2014); Bharadhirajan *et al.* (2015); Khan (2015); Mahesh & Saravanakumar (2015)
Photopectoralis bindus (Valenciennes, 1835)	-	-	+	-	-	-	-	-	Marine; Brackish	Minor commercial	11.0 cm	NE	Bharadhirajan *et al.* (2015); Khan (2015)

Order, Family, Scientific name	Distribution of fishes in the Estuaries of Tamil Nadu								Environment	Human use oe fishery status	Maximum size (TL)	IUCN Status	References
	I	II	III	IV	V	VI	VII	VIII					
Secutor insidiator (Bloch, 1787)	+	+	+	+	+	-	-	-	Marine; Brackish	Commercial	11.3 cm SL	NE	Remadevi *et al.* (2004); Ramanujam & Anbarasan (2008); Ramanujam *et al.* (2014); Bharadhirajan *et al.* (2015); Khan (2015); Mahesh & Saravanakumar (2015); Pavinkumar *et al.* (2015)
Secutor ruconius (Hamilton, 1822)	+	+	+	-	+	-	-	-	Marine; Freshwater; Brackish	Minor commercial	8.0 cm	NE	Remadevi *et al.* (2004); Ramanujam & Anbarasan (2008); Ramanujam *et al.* (2014); Khan (2015); Mahesh & Saravanakumar (2015)
Family: Lethrinidae													
Lethrinus lentjan (Lacepède, 1802)	-	-	-	+	-	-	-	-	Marine; Brackish	Highly commercial	52.0 cm	NE	Pavinkumar *et al.* (2015)
Lethrinus nebulosus (Forsskål, 1775)	-	+	+	-	-	-	-	-	Marine; Brackish	Highly commercial; Aquaculture; Gamefish	87.0 cm	NE	Remadevi *et al.* (2004); Khan (2015)
Family: Latidae													

Order, Family, Scientific name	Distribution of fishes in the Estuaries of Tamil Nadu								Environment	Human use oe fishery status	Maximum size (TL)	IUCN Status	References
	I	II	III	IV	V	VI	VII	VIII					
Lates calcarifer (Bloch, 1790)	+	-	+	+	-	+	-	-	Marine; Freshwater; Brackish	Highly commercial; Aquaculture; Gamefish; Ornamental	200 cm	NE	Johnson & Selvaraj (2008); Kumaran *et al.* (2012); Kannappan & Karthikeyan (2013); Ramanujam *et al.* (2014); Bharadhirajan *et al.* (2015); Khan (2015); Mahesh & Saravanakumar (2015); Pavinkumar *et al.* (2015)
Family: Lobotidae													
Lobotes surinamensis (Bloch, 1790)	-	-	+	-	-	-	-	-	Marine; Brackish	Commercial; Gamefish; Ornamental	110 cm	NE	Khan (2015)
Family: Lutjanidae													
Lutjanus argentimaculatus (Forsskål, 1775)	-	+	+	+	+	-	-	-	Marine; Freshwater; Brackish	Commercial; Aquaculture; Gamefish	150 cm	NE	Remadevi *et al.* (2004); Ramanujam & Anbarasan (2008); Bharadhirajan *et al.* (2015); Khan (2015); Mahesh & Saravanakumar (2015); Pavinkumar *et al.* (2015)
Lutjanus fulviflamma (Forsskål, 1775)	-	-	+	+	+	-	+	-	Marine; Brackish	Commercial; Gamefish; Ornamental	35.0 cm	NE	Ramanujam & Anbarasan (2008); Sukumaran *et al.* (2014); Bharadhirajan *et al.* (2015); Khan (2015); Pavinkumar *et al.* (2015)

Order, Family, Scientific name	Distribution of fishes in the Estuaries of Tamil Nadu								Environment	Human use oe fishery status	Maximum size (TL)	IUCN Status	References
	I	II	III	IV	V	VI	VII	VIII					
Lutjanus gibbus (Forsskål, 1775)	-	-	-	+	-	-	-	-	Marine	Commercial; Gamefish; Ornamental	50.0 cm	NE	Kannappan & Karthikeyan (2013)
Lutjanus johnii (Bloch, 1792)	-	-	+	+	-	-	-	-	Marine; Brackish	Commercial; Aquaculture; Gamefish	97.0 cm	NE	Khan (2015); Pavinkumar *et al.* (2015)
Lutjanus kasmira (Forsskål, 1775)	-	+	-	-	-	-	-	-	Marine	Commercial; Gamefish; Ornamental	40.0 cm	NE	Remadevi *et al.* (2004)
Lutjanus russellii (Bleeker, 1849)	-	-	+	-	-	-	-	-	Marine; Brackish	Commercial; Aquaculture	50.0 cm	NE	Khan (2015); Mahesh & Saravanakumar (2015)
Family: Menidae													
Mene maculata (Bloch & Schneider, 1801)	-	-	-	+	-	-	-	-	Marine; Brackish	Commercial	30.0 cm	NE	Kannappan & Karthikeyan (2013)
Family: Monodactylidae													
Monodactylus argenteus (Linnaeus, 1758)	-	-	-	+	-	-	-	-	Marine; Freshwater; Brackish	Minor commercial; Ornamental	27.0 cm SL	NE	Pavinkumar *et al.* (2015)
Family: Mullidae													
Parupeneus indicus (Shaw, 1803)	-	-	-	-	-	+	-	-	Marine; Brackish	Commercial; Gamefish	45.0 cm	NE	Kumaran *et al.* (2012)
Upeneus moluccensis (Bleeker, 1855)	-	-	-	+	-	-	-	-	Marine; Brackish	Commercial	22.0 cm	NE	Pavinkumar *et al.* (2015)

Order, Family, Scientific name	Distribution of fishes in the Estuaries of Tamil Nadu								Environment	Human use oe fishery status	Maximum size (TL)	IUCN Status	References
	I	II	III	IV	V	VI	VII	VIII					
Upeneus sulphureus Cuvier, 1829	+	+	+	-	-	-	-	-	Marine; Brackish	Minor commercial; Ornamental	23.0 cm	NE	Remadevi *et al.* (2004); Ramanujam *et al.* (2014); Khan (2015); Mahesh & Saravanakumar (2015)
Upeneus vittatus (Forsskål, 1775)	-	-	-	-	-	+	-	-	Marine; Brackish	Commercial; Ornamental	28.0 cm	NE	Kumaran *et al.* (2012)
Family: Nandidae													
Nandus nandus (Hamilton, 1822)	-	+	-	-	-	-	-	-	Freshwater; Brackish	Commercial; Ornamental	20.0 cm	LC	Remadevi *et al.* (2004)
Family: Osphronemidae													
Pseudosphromenus cupanus (Cuvier, 1831)	-	-	-	-	+	-	-	-	Freshwater; Brackish	Ornamental	7.5 cm	LC	Ramanujam & Anbarasan (2008)
Family: Polynemidae													
Eleutheronema tetradactylum (Shaw, 1804)	-	-	+	-	-	-	-	-	Marine; Freshwater; Brackish	Highly commercial; Aquaculture	200 cm	NE	Bharadhirajan *et al.* (2015); Khan (2015)
Polydactylus plebeius (Broussonet, 1782)	+	-	-	-	-	-	-	-	Marine; Brackish	Commercial; Aquaculture; Gamefish	45.0 cm SL	NE	Ramanujam *et al.* (2014)
Family: Pomacentridae													
Abudefduf septemfasciatus (Cuvier, 1830)	-	-	-	+	-	-	-	-	Marine	Minor commercial	23.0 cm	NE	Pavinkumar *et al.* (2015)
Family: Scaridae													

Order, Family, Scientific name	Distribution of fishes in the Estuaries of Tamil Nadu								Environment	Human use oe fishery status	Maximum size (TL)	IUCN Status	References
	I	II	III	IV	V	VI	VII	VIII					
Scarus ghobban Forsskål, 1775	-	-	-	+	-	-	-	-	Marine; Brackish	Commercial; Ornamental	90.0 cm	LC	Pavinkumar *et al.* (2015)
Family: Scatophagidae													
Scatophagus argus (Linnaeus, 1766)	+	-	+	+	+	-	+	-	Marine; Freshwater; Brackish	Minor commercial; Aquaculture; Ornamental	38.0 cm	LC	Ramanujam & Anbarasan (2008); Ramanujam *et al.* (2014); Sukumaran *et al.* (2014); Bharadhirajan *et al.* (2015); Khan (2015); Mahesh & Saravanakumar (2015); Pavinkumar *et al.* (2015)
Family: Sciaenidae													
Daysciaena albida (Cuvier, 1830)	-	+	-	-	-	-	-	-	Marine; Brackish	Commercial	90.0 cm SL	NE	Remadevi *et al.* (2004)
Dendrophysa russelii (Cuvier, 1829)	-	-	+	-	-	-	-	-	Marine; Freshwater; Brackish	Minor commercial	25.0 cm SL	NE	Khan (2015)
Johnius carutta Bloch, 1793	-	-	+	-	-	-	-	-	Marine; Freshwater; Brackish	Minor commercial	30.0 cm SL	NE	Khan (2015)
Kathala axillaris (Cuvier, 1830)	-	-	+	-	-	-	-	-	Marine	Commercial	27.0 cm	NE	Khan (2015)
Nibea maculata (Bloch & Schneider, 1801)	-	+	-	-	-	-	-	-	Marine	Commercial	30.0 cm	NE	Remadevi *et al.* (2004)
Family: Scombridae													

Order, Family, Scientific name	Distribution of fishes in the Estuaries of Tamil Nadu								Environment	Human use oe fishery status	Maximum size (TL)	IUCN Status	References
	I	II	III	IV	V	VI	VII	VIII					
Rastrelliger kanagurta (Cuvier, 1816)	-	-	-	-	-	+	-	-	Marine	Highly commercial; Gamefish; Bait	38.0 cm	DD	Kumaran *et al.* (2012)
Scomberomorus commerson (Lacepède, 1800)	-	-	+	-	-	-	-	-	Marine	Highly commercial; Gamefish	240 cm	NT	Khan (2015)
Scomberomorus guttatus (Bloch & Schneider, 1801)	-	-	-	-	-	+	-	-	Marine; Brackish	Highly commercial; Gamefish	76.0 cm FL	DD	Kumaran *et al.* (2012)
Family: Serranidae													
Epinephelus bleekeri (Vaillant, 1878)	-	-	-	+	-	-	-	-	Marine	Minor commercial; Aquaculture	76.0 cm	NT	Bharadhirajan *et al.* (2015)
Epinephelus diacanthus (Valenciennes, 1828)	-	-	-	+	-	-	-	-	Marine	Commercial	55.0 cm	NT	Pavinkumar *et al.* (2015)
Epinephelus malabaricus (Bloch & Schneider, 1801)	-	+	-	+	-	-	-	-	Marine; Brackish	Highly commercial; Aquaculture; Gamefish	234 cm	NT	Remadevi *et al.* (2004); Pavinkumar *et al.* (2015)
Epinephelus morrhua (Valenciennes, 1833)	-	+	-	-	-	-	-	-	Marine	Minor commercial; Gamefish	90.0 cm	LC	Remadevi *et al.* (2004)

Order, Family, Scientific name	Distribution of fishes in the Estuaries of Tamil Nadu								Environment	Human use oe fishery status	Maximum size (TL)	IUCN Status	References
	I	II	III	IV	V	VI	VII	VIII					
Epinephelus tauvina (Forsskål, 1775)	-	-	+	+	+	-	-	-	Marine	Minor commercial; Aquaculture; Gamefish	100.0 cm	DD	Ramanujam & Anbarasan (2008); Bharadhirajan *et al.* (2015); Khan (2015); Mahesh & Saravanakumar (2015); Pavinkumar *et al.* (2015)
Family: Siganidae													
Siganus canaliculatus (Park, 1797)	-	-	+	+	-	-	-	-	Marine; Brackish	Commercial; Aquaculture	30.0 cm	NE	Bharadhirajan *et al.* (2015); Khan (2015); Mahesh & Saravanakumar (2015); Pavinkumar *et al.* (2015)
Siganus javus (Linnaeus, 1766)	+	-	+	+	+	-	+	-	Marine; Brackish	Commercial; Ornamental	53.0 cm	NE	Ramanujam & Anbarasan (2008); Ramanujam *et al.* (2014); Sukumaran *et al.* (2014); Bharadhirajan *et al.* (2015); Khan (2015); Mahesh & Saravanakumar (2015); Pavinkumar *et al.* (2015)
Siganus lineatus (Valenciennes, 1835)	-	-	-	+	-	-	-	-	Marine	Commercial; Ornamental	43.0 cm	NE	Pavinkumar *et al.* (2015)
Family: Sillaginidae													

Order, Family, Scientific name	Distribution of fishes in the Estuaries of Tamil Nadu								Environment	Human use oe fishery status	Maximum size (TL)	IUCN Status	References
	I	II	III	IV	V	VI	VII	VIII					
Sillago sihama (Forsskål, 1775)	+	+	+	+	+	-	+	+	Marine; Brackish	Commercial; Aquaculture	31.0 cm SL	NE	Remadevi *et al.* (2004); Johnson & Selvaraj (2008); Ramanujam & Anbarasan (2008); Kannappan & Karthikeyan (2013); Ramanujam *et al.* (2014); Sukumaran *et al.* (2014); Bharadhirajan *et al.* (2015); Khan (2015); Pavinkumar *et al.* (2015); Raju *et al.* (2015)
Sillago vincenti McKay, 1980	-	+	+	+	-	-	-	-	Marine; Brackish	Minor commercial; Aquaculture	30.0 cm SL	NE	Remadevi *et al.* (2004); Mahesh & Saravanakumar (2015); Pavinkumar *et al.* (2015)
Family: Sparidae													
Acanthopagrus berda (Forsskål, 1775)	-	+	-	-	-	-	-	-	Marine; Freshwater; Brackish	Commercial; Gamefish	90.0 cm	NE	Remadevi *et al.* (2004)
Acanthopagrus sivicolus Akazaki, 1962	-	-	-	+	-	-	-	-	Marine; Brackish	Commercial; Aquaculture	45.0 cm SL	NE	Pavinkumar *et al.* (2015)
Crenidens crenidens (Forsskål, 1775)	-	+	-	-	-	-	-	-	Marine	Commercial; Bait	30.0 cm	NE	Remadevi *et al.* (2004)
Rhabdosargus sarba (Forsskål, 1775)	-	+	-	-	-	-	-	-	Marine; Brackish	Commercial; Aquaculture; Gamefish	80.0 cm	NE	Remadevi *et al.* (2004)
Family: Sphyraenidae													

Order, Family, Scientific name	Distribution of fishes in the Estuaries of Tamil Nadu								Environment	Human use oe fishery status	Maximum size (TL)	IUCN Status	References
	I	II	III	IV	V	VI	VII	VIII					
Sphyraena barracuda (Edwards, 1771)	-	-	+	-	-	-	-	-	Marine; Brackish	Minor commercial; Gamefish; Ornamental	200 cm	NE	Khan (2015)
Sphyraena jello Cuvier, 1829	+	-	+	+	-	-	-	-	Marine; Brackish	Commercial; Gamefish	150 cm	NE	Ramanujam *et al.* (2014); Bharadhirajan *et al.* (2015); Khan (2015); Mahesh & Saravanakumar (2015); Pavinkumar *et al.* (2015)
Family: Stromatidae													
Pampus argenteus (Euphrasen, 1788)	-	-	-	-	-	+	-	-	Marine	Highly commercial	60.0 cm SL	NE	Kumaran *et al.* (2012)
Pampus chinensis (Euphrasen, 1788)	-	-	+	-	-	-	-	-	Marine; Brackish	Minor commercial	40.0 cm	NE	Khan (2015)
Family: Teraponidae													
Pelates quadrilineatus (Bloch, 1790)	-	-	+	+	-	-	-	-	Marine; Brackish	Minor commercial	30.0 cm	NE	Mahesh & Saravanakumar (2015); Pavinkumar *et al.* (2015)
Terapon jarbua (Forsskål, 1775)	+	+	+	+	+	-	-	-	Marine; Freshwater; Brackish	Minor commercial; Aquaculture	36.0 cm	LC	Remadevi *et al.* (2004); Johnson & Selvaraj (2008); Ramanujam & Anbarasan (2008); Kannappan & Karthikeyan (2013); Ramanujam *et al.* (2014); Bharadhirajan *et al.* (2015); Khan (2015); Mahesh & Saravanakumar (2015); Pavinkumar *et al.* (2015)

Order, Family, Scientific name	Distribution of fishes in the Estuaries of Tamil Nadu								Environment	Human use oe fishery status	Maximum size (TL)	IUCN Status	References
	I	II	III	IV	V	VI	VII	VIII					
Terapon puta Cuvier, 1829	+	+	+	+	+	-	-	-	Marine; Freshwater; Brackish	Minor commercial	16.0 cm	NE	Remadevi *et al.* (2004); Ramanujam & Anbarasan (2008); Ramanujam *et al.* (2014); Bharadhirajan *et al.* (2015); Khan (2015); Mahesh & Saravanakumar (2015); Pavinkumar *et al.* (2015)
Terapon theraps Cuvier, 1829	-	+	-	-	-	-	-	-	Marine; Freshwater; Brackish	Minor commercial	30.0 cm SL	LC	Remadevi *et al.* (2004)
Family: Trichiuridae													
Lepturacanthus savala (Cuvier, 1829)	-	-	-	-	-	+	-	-	Marine; Brackish	Commercial	100.0 cm SL	NE	Kumaran *et al.* (2012)
Trichiurus lepturus Linnaeus, 1758	-	-	-	-	-	+	-	-	Marine; Brackish	Highly commercial; Gamefish	234 cm	NE	Kumaran *et al.* (2012)
Family: Uranoscopidae													
Uranoscopus guttatus Cuvier, 1829	-	+	-	-	-	-	-	-	Marine	Fisheries of no interest	20.0 cm	NE	Remadevi *et al.* (2004)
Order: Pleuronectiforms													
Family: Cynoglossidae													
Cynoglossus arel (Bloch & Schneider, 1801)	-	-	+	+	-	-	-	-	Marine; Brackish	Commercial	40.0 cm	NE	Bharadhirajan *et al.* (2015); Mahesh & Saravanakumar (2015); Khan (2015); Pavinkumar *et al.* (2015)

Order, Family, Scientific name	Distribution of fishes in the Estuaries of Tamil Nadu								Environment	Human use oe fishery status	Maximum size (TL)	IUCN Status	References
	I	II	III	IV	V	VI	VII	VIII					
Cynoglossus lida (Bleeker, 1851)	-	-	+	-	-	-	-	-	Marine	Minor commercial	21.3 cm SL	NE	Khan (2015)
Cynoglossus lingua Hamilton, 1822	-	-	-	+	-	-	-	-	Marine; Freshwater; Brackish	Commercial	45.0 cm	NE	Kannappan & Karthikeyan (2013)
Cynoglossus macrolepidotus (Bleeker, 1851)	-	-	-	-	-	+	-	-	Marine	Fisheries of no interest	29.0 cm	NE	Kumaran *et al.* (2012)
Cynoglossus puncticeps (Richardson, 1846)	-	-	+	-	-	-	-	-	Marine; Freshwater; Brackish	Commercial	35.0 cm SL	NE	Bharadhirajan *et al.* (2015); Khan (2015)
Family: Soleidae													
Brachirus orientalis (Bloch & Schneider, 1801)	-	+	-	+	+	-	-	-	Marine; Freshwater; Brackish	Commercial	30.0 cm SL	NE	Remadevi *et al.* (2004); Ramanujam & Anbarasan (2008); Pavinkumar *et al.* (2015)
Pseudorhombus arsius (Hamilton, 1822)	-	-	-	-	+	-	-	-	Marine; Brackish	Commercial; Gamefish	45.0 cm	NE	Ramanujam & Anbarasan (2008)
Order: Scorpaeniformes													
Family: Platycephalidae													
Grammoplites scaber (Linnaeus, 1758)	-	+	-	-	-	-	-	-	Marine; Brackish	Minor commercial	30.0 cm	NE	Remadevi *et al.* (2004)

Order, Family, Scientific name	Distribution of fishes in the Estuaries of Tamil Nadu								Environment	Human use oe fishery status	Maximum size (TL)	IUCN Status	References
	I	II	III	IV	V	VI	VII	VIII					
Platycephalus indicus (Linnaeus, 1758)	-	+	+	+	+	-	-	-	Marine; Brackish	Commercial; Aquaculture; Gamefish	100.0 cm	DD	Remadevi *et al.* (2004); Ramanujam & Anbarasan (2008); Bharadhirajan *et al.* (2015); Mahesh & Saravanakumar (2015); Pavinkumar *et al.* (2015)
Order: Siluriformes													
Family: Ariidae													
Arius arius (Hamilton, 1822)	+	+	+	+	+	-	-	-	Marine; Brackish	Commercial	40.0 cm SL	LC	Remadevi *et al.* (2004); Ramanujam & Anbarasan (2008); Ramanujam *et al.* (2014); Bharadhirajan *et al.* (2015); Khan (2015); Pavinkumar *et al.* (2015)
Arius jella Day, 1877	+	-	-	+	-	-	-	-	Marine; Brackish	Commercial	30.0 cm	NE	Ramanujam *et al.* (2014); Pavinkumar *et al.* (2015)
Arius maculatus (Thunberg, 1792)	+	+	+	+	+	-	+	-	Marine; Freshwater; Brackish	Commercial	80.0 cm	NE	Remadevi *et al.* (2004); Ramanujam & Anbarasan (2008); Kannappan & Karthikeyan (2013); Ramanujam *et al.* (2014); Sukumaran *et al.* (2014); Mahesh & Saravanakumar (2015); Pavinkumar *et al.* (2015)

Order, Family, Scientific name	Distribution of fishes in the Estuaries of Tamil Nadu								Environment	Human use oe fishery status	Maximum size (TL)	IUCN Status	References
	I	II	III	IV	V	VI	VII	VIII					
Arius subrostratus Valenciennes, 1840	-	-	+	+	+	-	-	-	Marine; Brackish	Commercial	39.5 cm	NE	Ramanujam & Anbarasan (2008); Bharadhirajan *et al.* (2015); Khan (2015); Pavinkumar *et al.* (2015)
Nemapteryx caelata (Valenciennes, 1840)	-	+	-	-	-	-	-	-	Marine; Brackish	Commercial	45.0 cm	NE	Remadevi *et al.* (2004)
Netuma thalassina (Rüppell, 1837)	-	-	-	+	-	+	-	-	Marine; Freshwater; Brackish	Commercial; Gamefish	185 cm	NE	Kumaran *et al.* (2012); Kannappan & Karthikeyan (2013)
Plicofollis dussumieri (Valenciennes, 1840)	-	-	-	+	-	-	-	-	Marine; Freshwater; Brackish	Commercial	62.0 cm SL	LC	Kannappan & Karthikeyan (2013)
Sciades sona (Hamilton, 1822)	-	-	-	+	-	-	+	-	Marine; Brackish	Commercial	92.0 cm	NE	Sukumaran *et al.* (2014)
Family: Bagridae													
Mystus gulio (Hamilton, 1822)	+	+	-	+	+	-	+	+	Freshwater; Brackish	Commercial	46.0 cm	LC	Remadevi *et al.* (2004); Ramanujam & Anbarasan (2008); Ramanujam *et al.* (2014); Sukumaran *et al.* (2014); Pavinkumar *et al.* (2015); Raju *et al.* (2015)
Mystus vittatus (Bloch, 1794)	-	-	-	-	+	-	-	-	Freshwater; Brackish	Minor commercial; Ornamental	21.0 cm SL	LC	Ramanujam & Anbarasan (2008)
Family: Clariidae													

Order, Family, Scientific name	Distribution of fishes in the Estuaries of Tamil Nadu								Environment	Human use oe fishery status	Maximum size (TL)	IUCN Status	References
	I	II	III	IV	V	VI	VII	VIII					
Clarias batrachus (Linnaeus, 1758)	-	-	-	+	-	-	-	-	Freshwater; Brackish	Commercial; Aquaculture; Ornamental	47.0 cm	LC	Pavinkumar *et al.* (2015)
Family: Heteropneustidae													
Heteropneustes fossilis (Bloch, 1794)	-	-	-	-	-	+	-	-	Freshwater; Brackish	Highly commercial; Aquaculture; Ornamental	30.0 cm	LC	Kumaran *et al.* (2012)
Family: Plotosidae													
Plotosus canius Hamilton, 1822	-	+	+	-	+	-	+	-	Marine; Freshwater; Brackish	Commercial	150 cm	NE	Remadevi *et al.* (2004); Ramanujam & Anbarasan (2008); Sukumaran *et al.* (2014); Bharadhirajan *et al.* (2015); Khan (2015); Mahesh & Saravanakumar (2015)
Family: Siluridae													
Wallago attu (Bloch & Schneider, 1801)	-	-	+	-	-	-	-	-	Freshwater; Brackish	Commercial; Gamefish	240 cm	NT	Khan (2015)
Order: Synbranchiformes													
Family: Mastacembelidae													

Order, Family, Scientific name	Distribution of fishes in the Estuaries of Tamil Nadu								Environment	Human use oe fishery status	Maximum size (TL)	IUCN Status	References
	I	II	III	IV	V	VI	VII	VIII					
Macrognathus aral (Bloch & Schneider, 1801)	-	-	-	-	+	-	-	-	Freshwater; Brackish	Commercial	63.5 cm	LC	Ramanujam & Anbarasan (2008)
Macrognathus pancalus Hamilton, 1822	-	-	-	-	+	-	-	-	Freshwater; Brackish	Minor commercial	18.0 cm	LC	Ramanujam & Anbarasan (2008)
Order: Tetrodotiformes													
Family: Tetrodontidae													
Arothron leopardus (Day, 1878)	-	-	-	+	-	-	-	-	Marine	Fisheries of no interest	15.0 cm	NE	Pavinkumar *et al.* (2015)
Tetraodon cutcutia (Hamilton, 1822)	-	+	-	-	-	-	-	-	Freshwater; Brackish	Fisheries of no interest	15.0 cm	LC	Remadevi *et al.* (2004)
Tetraodon fluviatilis Hamilton, 1822	-	-	-	-	+	-	-	-	Freshwater; Brackish	Ornamental	17.0 cm	NE	Ramanujam & Anbarasan (2008)
Family: Triacanthidae													
Chelonodon patoca (Hamilton, 1822)	-	+	-	-	-	-	-	-	Marine; Freshwater; Brackish	Minor commercial	38.0 cm SL	NE	Remadevi *et al.* (2004)
Lagocephalus lunaris (Bloch & Schneider, 1801)	-	+	-	-	-	-	-	-	Marine; Brackish	Ornamental	45.0 cm SL	NE	Remadevi *et al.* (2004)
Takifugu oblongus (Bloch, 1786)	-	+	-	-	-	-	-	-	Marine; Brackish	Fisheries of no interest	40.0 cm	NE	Remadevi *et al.* (2004)

Order, Family, Scientific name	Distribution of fishes in the Estuaries of Tamil Nadu								Environment	Human use oe fishery status	Maximum size (TL)	IUCN Status	References
	I	II	III	IV	V	VI	VII	VIII					
Triacanthus biaculeatus (Bloch, 1786)	+	+	-	+	+	-	-	-	Marine; Brackish	Minor commercial	30.0 cm	NE	Remadevi *et al.* (2004); Ramanujam & Anbarasan (2008); Ramanujam *et al.* (2014); Pavinkumar *et al.* (2015)

Distribution of fishes in the Estuaries of Tamil Nadu:

I: Adyar; II: Pulicat lake; III: Coleroon-Pichavaram-Vellar; IV: Manakudy; V: Yedayanthittu; VI: Giriyampeta; VII: Muthupet; VIII: Arasalar

IUCN Red list Status:

VU: Vulnerable; NT: Near Threatened; LR/ lc: Lower Risk-least concern; LC: Least Concern; DD: Data Deficient; NE: Not Evaluated

Table 2. Summary of the estuarine fishes found in Tamil Nadu.

Taxa	Chondrichthyes	Osteichthyes	Total
Order	3	14	17
Families	3	172	75
Genera	3	151	154
Species	3	138	241

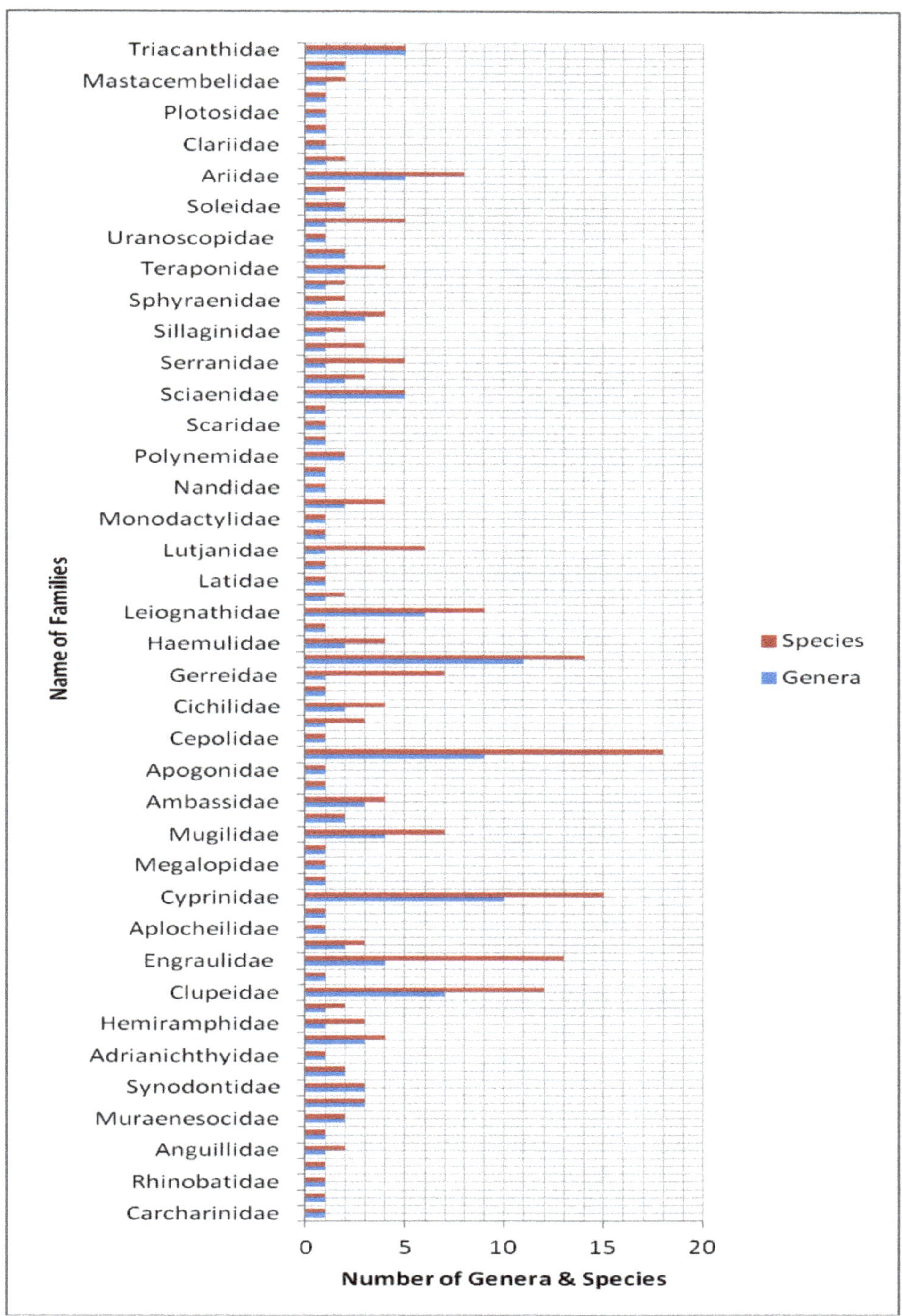
Triacanthidae
Mastacembelidae
Plotosidae
Clariidae
Ariidae
Soleidae
Uranoscopidae
Teraponidae
Sphyraenidae
Sillaginidae
Serranidae
Sciaenidae
Scaridae
Polynemidae
Nandidae
Monodactylidae
Lutjanidae
Latidae
Leiognathidae
Haemulidae
Gerreidae
Cichilidae
Cepolidae
Apogonidae
Ambassidae
Mugilidae
Megalopidae
Cyprinidae
Aplocheilidae
Engraulidae
Clupeidae
Hemiramphidae
Adrianichthyidae
Synodontidae
Muraenesocidae
Anguillidae
Rhinobatidae
Carcharinidae
Name of Families
0
5
10
15
20
Number of Genera & Species
Species
Genera

Figure 1. Family wise representation of estuarine fishes of Tamil Nadu.

Figure 2. Fishes recorded by various authors as well as total number of fishes recorded in the estuaries of Tamil Nadu.

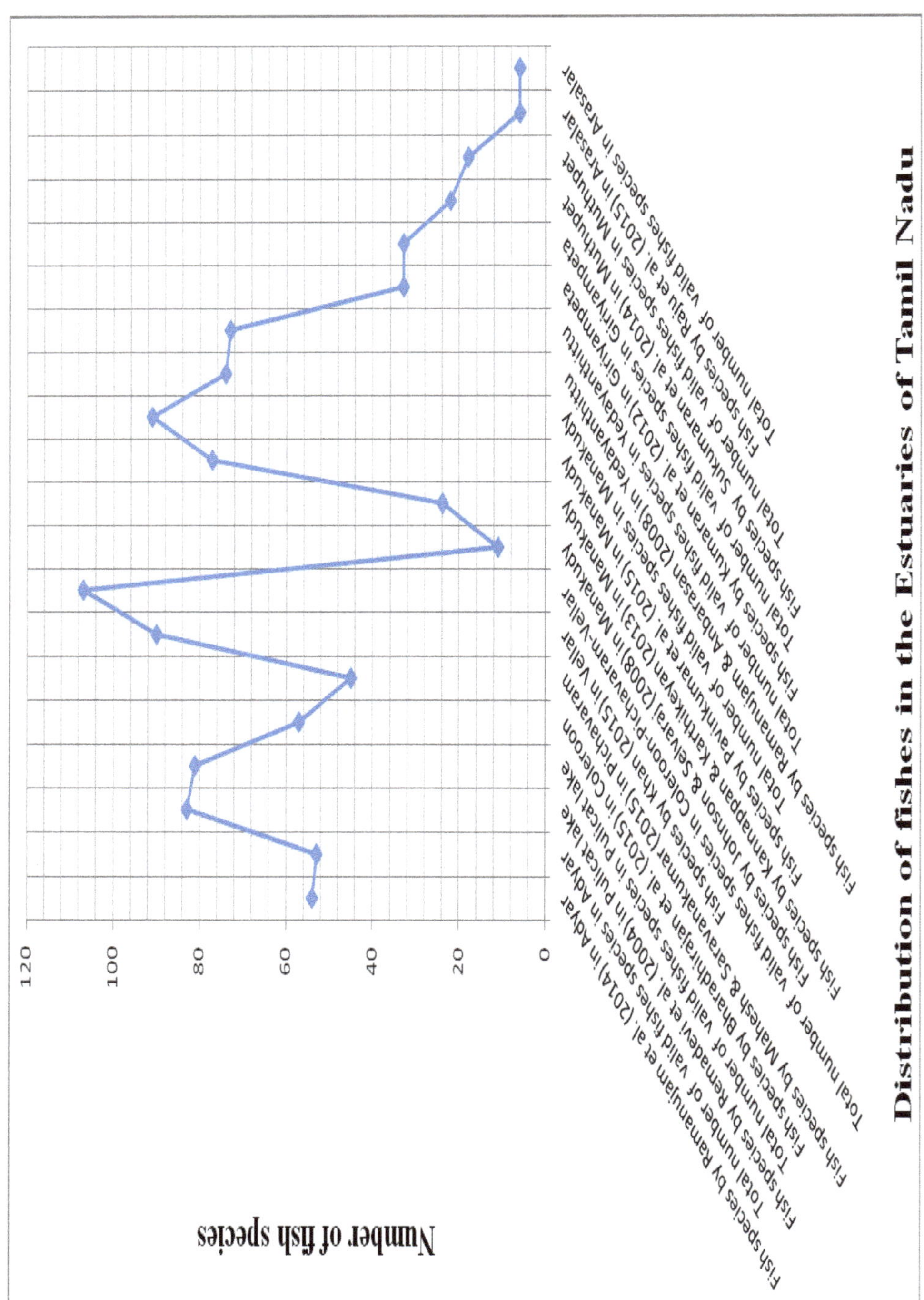

7

FINFISH DIVERSITY OF VEMBANAD LAKE IN THE PANANGAD-KUMBALAM REGION OF KOCHI, KERALA

***C.P. Ansar*[1], *H.S. Mogalekar*[2], *N.N. Raman*[3], *K.V. Jayachandran*[3] and *K. Dinesh*[3]**

[1] *Regional Agriculture Research Station, Kumarokom- 686 563, Kerala, India*

[2] *Fisheries College and Research Institute, Thoothukudi – 628 008, Tamil Nadu, India.*

[3] *Kerala University of Fisheries and Ocean Studies, Kochi – 682 506, Kerala, India.*

INTRODUCTION

Vembanad Lake is second large brackishwater lake in the India and is among the most productive life-supporting coastal wetland in Kerala, having length of 96 km and surface area of 1512 km^2 (Krishna and Priyadarsanan, 2012). Six rivers bring freshwaters into Vembanad Lake and it has two permanent opening to the Arabian Sea one at Cochin and other at Azhikode. Vembanad Lake is a transitional ecotone between sea and land encompassing mangroves, mudflats, swamps and marshes. As these ecosystems provide a harsh environment, many species of finfish have found them to be an ideal place for spawning, development and growth during early life. Rich biodiversity and ecological value made Vembanad Lake to be identified as a Ramsar site in November 2002.

Decline in estuarine diversity as a result of overfishing, insufficient management and habitat degradation, reduces the chances of its sustainability (James, 1987). Extensive water and land remodeling efforts, destructive fishing methods and unprecedented growth of noxious aquatic weeds like Salvinia and Eichornia in Vembanad Lake has created an alarming situation in the area. Therefore, knowledge on the status and trends of backwater fisheries is the key to sound policy development, better decision-making and responsible fisheries management. In this sense, in present investigation an attempt has been made to list out the finfish diversity in the Vembanad Lake at Panangad-Kumbalam backwater.

MATERIALS AND METHODS

The present investigation (June 2012 to May 2013) scrutinized the status of finfish diversity of Vembanad Lake at Panangad-Kumbalam backwater. Panangad-Kumbalam backwater is located at northern extremity of Vembanad wetland ecosystem (Latitude 09.90990^0 to 09.90638^0 N and Longitude 76.31475^0 to 76.31514^0 E) in Ernakulam district of Kerala. Finfish sampling was carried out by cast nets (22 mm), dragnet (10 mm) and gillnet (25 mm) during low tide on monthly basis. After sorting, representative finfish fauna were identified up to species level (Day, 1889; Talwar and Jhingran, 1991; Jayaram, 2010).

RESULTS AND DISCUSSION

The primary aim of the study was to list out the finfish diversity as the backwater ecosystems located in coastal areas are subjected to increasing pressures from anthropogenic and climate factors. In total, 2231 individuals of finfishes were encountered during study period, comprising 38 species of finfishes belonging to 27 families 11 orders and 31 genera (Table 1). Based on the information collected on conservation status of all taxa, three species listed as near threatened, 18 as least concern, 2 as data deficient and 15 species have not yet been evaluated for their conservation status by IUCN Red list (Table 1). Distribution and abundance of fish diversity in Vembanad Lake has been extensively by Kurup and Samuel (1987); Kurup *et al.* (1989); Bijoy *et al.* (2012). 150 species of fish belonging to 100 genera and 56 families are known to occur in Vembanad Lake (Kurup and Samuel 1985). Krishna and Priyadarsanan, (2012) recorded 67 species of finfishes; whereas, Bijoy *et al.* (2012) 60 species of finfishes from Vembanad Lake. However, present record of 39 species is lower compared to all the above reports. The lower number of species in Panangad-Kumbalam backwater is understandable as the all the above reports covered large area of Vembanad Lake than the present study. According to Kurup *et al.* (1989) the fish fauna of Vembanad Lake was dominated by the marine migratory species (56%). During this study saline tolerant freshwater species dominates (15%) and the marine migrants are increased to 85%. Species richness was highly reduced in Vembanad Lake due to decreased waterscape, pollution, loss of mangroves, unsustainable fishing practices and heavy infestation of exotic weeds and fishes are identified as the major threats to the fishery of Vembanad Lake. The dominant taxa in Vembanad Lake were mostly similar. Family composition study revealed the representation of Cichlidae and Mugilidae was higher than other families. Similar dominance of Cichlidae and Mugilidae was indicated by Bijoy *et al.* (2012).

CONCLUSION

Recreational fishing may be organised in the Panangad-Kumbalam backwater as this region is endowed with food, ornamental and sport fishes.

REFERENCES

Bijoy, N.S., Jayachandran, P.R., Sreedevi, O.K. 2012. Temporal pattern of fish production in a microtidal tropical estuary in the south-west coast of India. Indian J. Fish., 59(2): 17-26.

Day, F. 1889. The Fauna of British India, Including, Ceylon and Burma, Fishes. Taylor and Francis. London. 1 (2): 548 pp.

James, E.J. 1987. Studies on estuarine dynamics on the southeast coast of India. Proc. of National Seminor on Estuarine Management. Trivandram. 76-82.

Jayaram, K.C. 2010. The Fresh water fishes of Indian Region. Second Edition. Narendra Publishing house, New Delhi. 616.

Krishna, K.K. and Priyadarsanan, D.R. 2012. Fish and Fisheries in Vembanad Lake. Consolidated report of Vembanad fish count 2008- 2011. CERC, Ashoka Trust for Research in Ecology and the Environment, 50pp.

Kurup, B. M., Sebastian, M.J., Sankaran, M.J. and Rabindranatha, P. 1989. Exploited fishery resource of Vembanad Lake. Final Report. Kuttanad Water Balance Study . p 142.

Kurup, B.M. and Samuel, C.T. 1985. Fish and fishery resources of the Vembanad Lake. In: Harvest and Post-Harvest Technology of Fishes Proceedings of Symposium on harvest and post-harvesting technology of fishes. Society of Fisheries Technologists (India), Kochi, p. 77-82.

Talwar, P.K. and Jhingran, A.G. 1991. Inland Fishes of India and Adjacent Countries. Vol 1 and 2, Oxford and IBH Publishing Co., New Delhi, 1–1158 pp.

Table 1 Check list of finfishes recorded in the Panangad-Kumbalam backwater with note on environment, human use, maximum size and IUCN status.

Order, Family, Scientific name	Environment	Human use	Maximum size (TL)	IUCN Status
Order: Anguilliforms				
Family: Anguillidae				
Anguilla bengalensis (Gray, 1831)	Marine; Freshwater; Brackish	Food; Gamefish	200 cm	NT
Order: Beloniformes				
Family: Belonidae				
Strongylura strongylura (van Hasselt, 1823)	Marine; Brackish	Food	40.0 cm SL	LC

Order, Family, Scientific name	Environment	Human use	Maximum size (TL)	IUCN Status
Xenentodon cancila (Hamilton, 1822)	Marine; Freshwater; Brackish	Food; Ornamental	40.0 cm	LC
Order: Clupeiformes				
Family: Clupeidae				
Anodontostoma chacunda (Hamilton, 1822)	Marine; Freshwater; Brackish	Food	22.0 cm SL	NE
Family: Engraulidae				
Stolephorus commersonnii Lacepède, 1803	Marine; Brackish	Food	10.0 cm SL	NE
Order: Cyprinodontiformes				
Family: Aplocheilidae				
Aplocheilus blockii Arnold, 1911	Freshwater; brackish	Ornamental	6.0 cm	LC
Aplocheilus lineatus (Valenciennes, 1846)	Freshwater; brackish	Ornamental	10.0 cm	LC
Order: Elopiformes				
Family: Elopidae				
Elops machnata (Forsskål, 1775)	Marine; Brackish	Food; Gamefish	118 cm FL	LC
Elops saurus Linnaeus, 1766	Marine; brackish	Food; Gamefish	100.0 cm	LC
Family: Megalopidae				
Megalops cyprinoides (Broussonet, 1782)	Marine; Freshwater; Brackish	Food; Gamefish	150 cm	DD
Order: Gonorhynchiformes				
Family: Chanidae				
Chanos chanos (Forsskål, 1775)	Marine; Freshwater; Brackish	Food; Gamefish	180 cm SL	NE
Order: Mugiliformes				
Family: Mugilidae				
Chelon macrolepis (Smith, 1846)	Marine; Freshwater; Brackish	Food; Gamefish	60.0 cm SL	LC
Chelon parsia (Hamilton, 1822)	Marine; Freshwater; Brackish	Food	16.0 cm	NE
Mugil cephalus Linnaeus, 1758	Marine; Freshwater; Brackish	Food; Gamefish	100.0 cm SL	LC

Order, Family, Scientific name	Environment	Human use	Maximum size (TL)	IUCN Status
Order: Perciformes				
Family: Anabantidae				
Anabas testudineus (Bloch, 1792)	Freshwater; Brackish	Food; Ornamental	25.0 cm	DD
Family: Ambassidae				
Ambassis ambassis (Lacepède, 1802)	Marine; Freshwater; Brackish	Ornamental	15.0 cm	LC
Ambassis ambassis (Lacepède, 1802)	Marine; freshwater; brackish	Ornamental	15.0 cm	LC
Family: Carangidae				
Scomberoides tol (Cuvier, 1832)	Marine; Brackish	Food; Gamefish	60.0 cm	NE
Family: Latidae				
Lates calcarifer (Bloch, 1790)	Marine; Freshwater; Brackish	Food; Gamefish	200 cm	NE
Family: Cichilidae				
Etroplus maculatus (Bloch, 1795)	Freshwater; Brackish	Food; Ornamental	8.0 cm	LC
Etroplus suratensis (Bloch, 1790)	Freshwater; Brackish	Food; Ornamental	40.0 cm	LC
Oreochromis mossambicus (Peters, 1852)	Freshwater; Brackish	Food	39.0 cm SL	NT
Family: Eleotridae				
Butis butis (Hamilton, 1822)	Marine; Freshwater; Brackish	Food	15.0 cm	LC
Family: Serranidae				
Epinephelus malabaricus (Bloch & Schneider, 1801)	Marine; Brackish	Food; Gamefish	234 cm	NT
Family: Gobiidae				
Glossogobius giuris (Hamilton, 1822)	Marine; Freshwater; Brackish	Food; Ornamental	50.0 cm SL	LC
Family: Leiognathidae				
Leiognathus brevirostris (Valenciennes, 1835)	Marine; Brackish	Food	13.5 cm	NE
Secutor insidiator (Bloch, 1787)	Marine; Brackish	Food	11.3 cm SL	NE

Order, Family, Scientific name	Environment	Human use	Maximum size (TL)	IUCN Status
Family: Lutjanidae				
Lutjanus argentimaculatus (Forsskål, 1775)	Marine; Freshwater; Brackish	Food; Gamefish	150 cm	NE
Lutjanus johnii (Bloch, 1792)	Marine; Brackish	Food; Gamefish	97.0 cm	NE
Family: Polynemidae				
Eleutheronema tetradactylum (Shaw, 1804)	Marine; Freshwater; Brackish	Food; Aquaculture	200 cm	NE
Family: Scatophagidae				
Scatophagus argus (Linnaeus, 1766)	Marine; Freshwater; Brackish	Food; Ornamental	38.0 cm	LC
Family: Sillaginidae				
Sillago sihama (Forsskål, 1775)	Marine; Brackish	Food; Aquaculture	31.0 cm SL	NE
Family: Teraponidae				
Terapon jarbua (Forsskål, 1775)	Marine; Freshwater; Brackish	Food	36.0 cm	LC
Order: Pleuronectiformes				
Family: Cynoglossidae				
Cynoglossus macrostomus Norman, 1928	Marine; Brackish	Food	17.3 cm	NE
Family: Psettodidae				
Psettodes erumei (Bloch & Schneider, 1801)	Marine	Food	64.0 cm	NE
Order: Siluriformes				
Family: Bagridae				
Sperata seenghala (Sykes, 1839)	Freshwater; brackish	Food; Gamefish	150 cm	LC
Mystus gulio (Hamilton, 1822)	Freshwater; Brackish	Food	46.0 cm	LC
Order: Tetrodotiformes				
Family: Tetrodontidae				
Arothron nigropunctatus (Bloch & Schneider, 1801)	Marine	Food	33.0 cm	NE

TL = Total Length; SL = Standard Length; IUCN - International Union for Conservation of Nature; NT = Near Threatened; LC = Least Concern; DD = Data Deficient; NE = Not Evaluated

Some of the exploited fishes of South India

Cirrhinus mrigala (Hamilton, 1822)

Catla catla (Hamilton,1822)

Labeo rohita (Hamilton, 1822)

Labeo gonius (Hamilton, 1822)

Labeo calbasu (Hamilton, 1822)

Puntius chola (Hamilton, 1822)

Systomus sarana (Hamilton, 1822)

Hypophthalmichthys molitrix (Valenciennes, 1844)

*Cyprinus carpio*Linnaeus, 1758

Rasbora daniconius (Hamilton, 1822)

Dawkinsia rohani (Rema Devi, Indra & Knight, 2010)

Dawkinsia arulius (Jerdon, 1849)

Mastacembelus armatus (Lacepède, 1800)

Anguilla bengalensis (Gray, 1831)

Anguilla bicolor McClelland, 1844

Conger cinereus Rüppell, 1830

Elops machnata (Forsskål, 1775)

Megalops cyprinoides (Broussonet, 1782)

Stolephorus commersonnii Lacepède, 1803

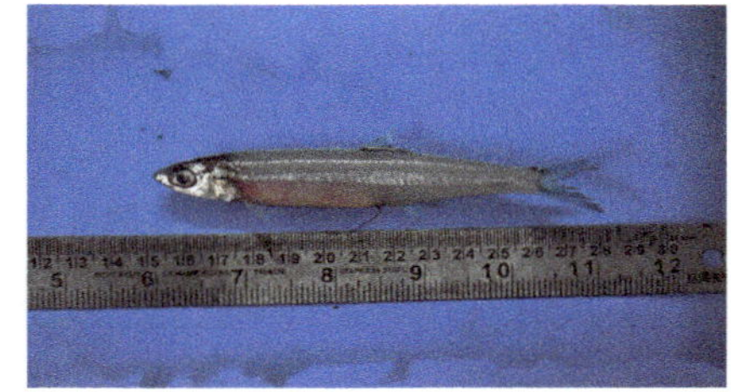

Stolephorus indicus (van Hasselt, 1823)

Thryssa vitrirostris (Gilchrist & Thompson, 1908)

Dussumieria acuta Valenciennes, 1847

Tenualosa ilisha (Hamilton, 1822)

Nematalosa nasus (Bloch, 1795)

Chanos chanos (Forsskål, 1775)

Clarias batrachus (Linnaeus, 1758)

Mystus bleekeri (Day, 1877)

Mystus gulio (Hamilton, 1822)

Mystus vittatus (Bloch, 1794)

Ompok bimaculatus (Bloch, 1794)

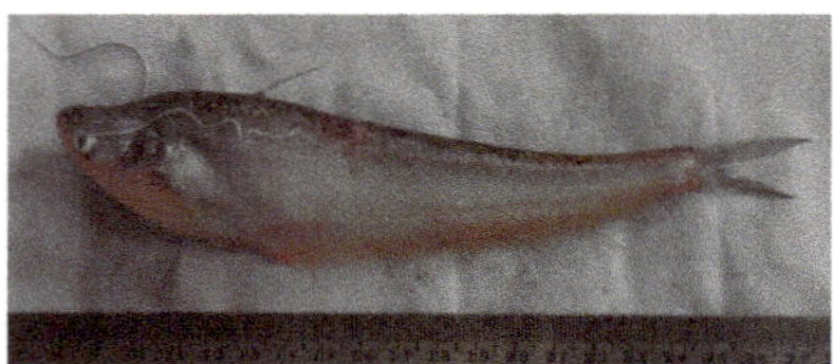

Ompok malabaricus (Valenciennes, 1840)

Arius arius (Hamilton, 1822)

Arius jella Day, 1877

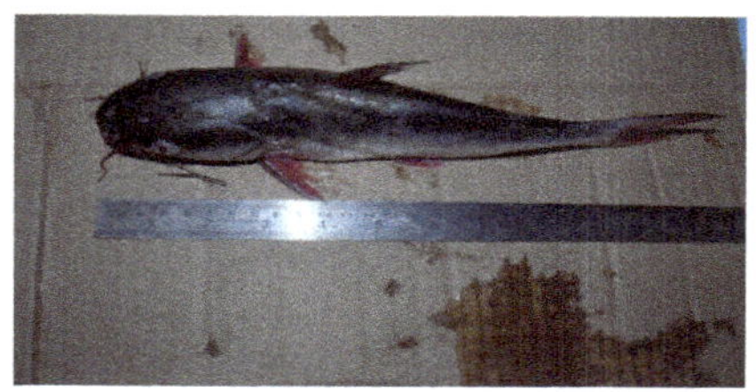

Arius maculatus (Thunberg, 1792)

Arius subrostratus Valenciennes, 1840

Moolgarda seheli (Forsskål, 1775)

Mugil cephalus Linnaeus, 1758

Chelon parsia (Hamilton, 1822)

Chelon subviridis (Valenciennes, 1836)

Chelon planiceps (Valenciennes, 1836)

Rhinomugil corsula (Hamilton, 1822)

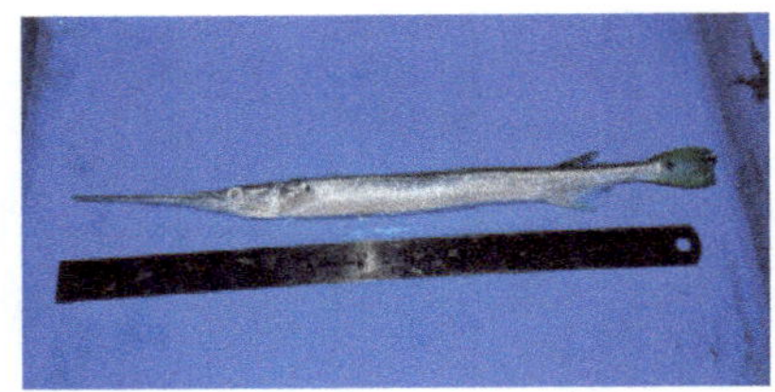

Strongylura strongylura (van Hasselt, 1823)

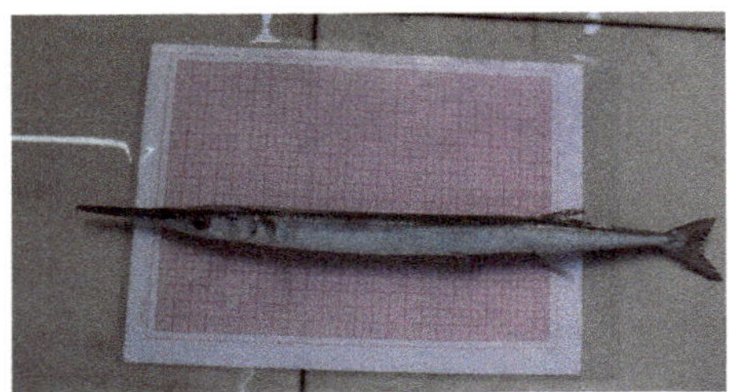

Tylosurus crocodilus (Péron & Lesueur, 1821)

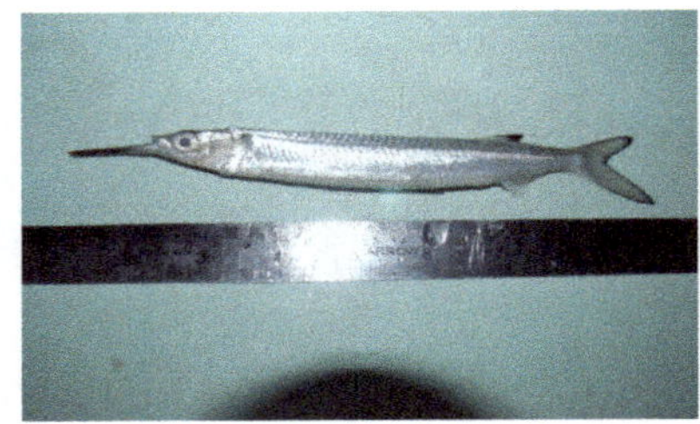

Hemiramphus marginatus (Forsskål, 1775)

Xenentodon cancila (Hamilton, 1822)

Anabas testudineus (Bloch, 1792)

Chanda nama Hamilton, 1822

Parambassis ranga (Hamilton, 1822)

Lates calcarifer (Bloch, 1790)

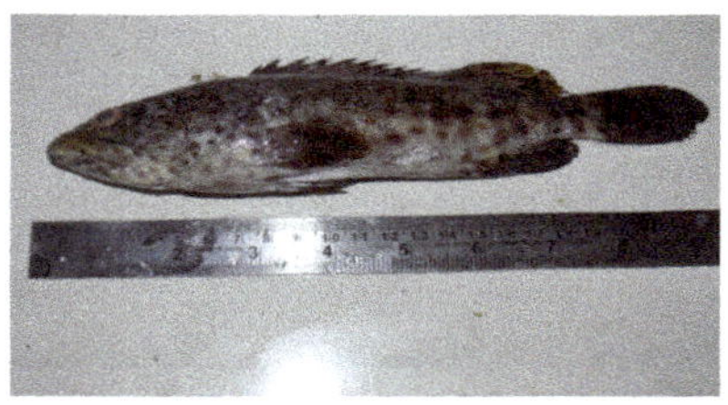

Epinephelus diacanthus (Valenciennes, 1828)

Epinephelus malabaricus (Bloch & Schneider, 1801)

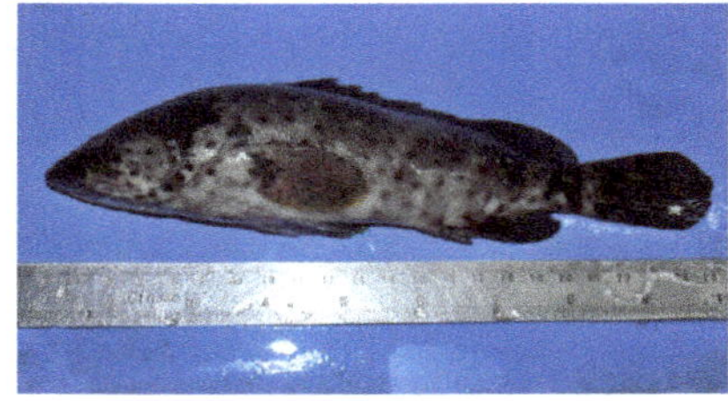

Epinephelus tauvina (Forsskål, 1775)

Sillago sihama (Forsskål, 1775)

Sillago vincenti McKay, 1980

Carangoides coeruleopinnatus (Rüppell, 1830)

Carangoides ferdau (Forsskål, 1775)

Caranx ignobilis (Forsskål, 1775)

Caranx sexfasciatus Quoy & Gaimard, 1825

Scomberoides commersonnianus Lacepède, 1801

Scomberoides tol (Cuvier, 1832)

Trachinotus africanus Smith, 1967

Leiognathus brevirostris (Valenciennes, 1835)

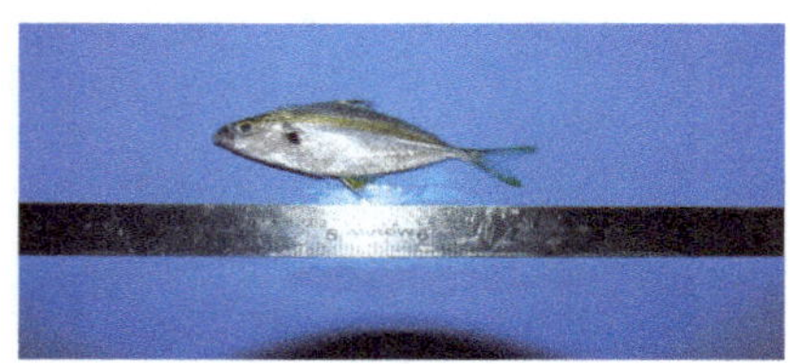

Karalla daura (Cuvier, 1829)

Eubleekeria splendens (Cuvier, 1829)

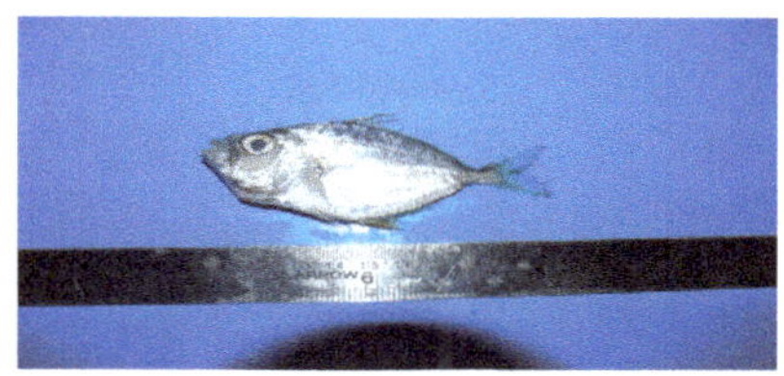

Secutor insidiator (Bloch, 1787)

Lutjanus argentimaculatus (Forsskål, 1775)

Lutjanus fulviflamma (Forsskål, 1775)

Lutjanus johnii (Bloch, 1792)

Gerres filamentosus Cuvier, 1829

Gerres oyena (Forsskål, 1775)

Gerres setifer (Hamilton, 1822)

Plectorhinchus gibbosus (Lacepède, 1802)

Acanthopagrus sivicolus Akazaki, 1962

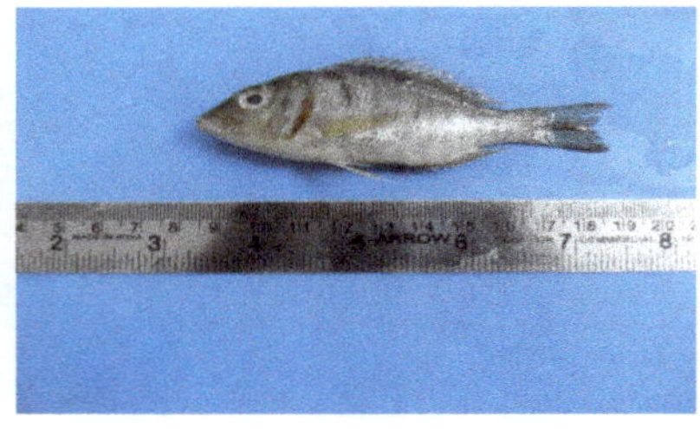

Lethrinus lentjan (Lacepède, 1802)

Upeneus moluccensis (Bleeker, 1855)

Monodactylus argenteus (Linnaeus, 1758)

Terapon jarbua (Forsskål, 1775)

Terapon puta Cuvier, 1829

Pelates quadrilineatus (Bloch, 1790)

Etroplus maculatus (Bloch, 1795)

Etroplus suratensis (Bloch, 1790)

Oreochromis mossambicus (Peters, 1852)

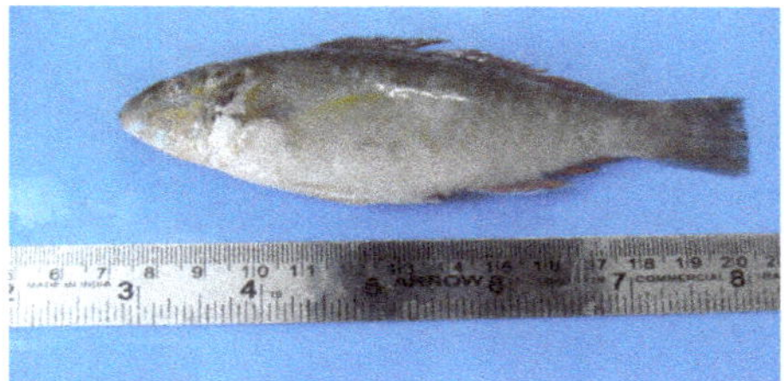

Scarus ghobban Forsskål, 1775

Acentrogobius ennorensis Menon & Rema Devi, 1980

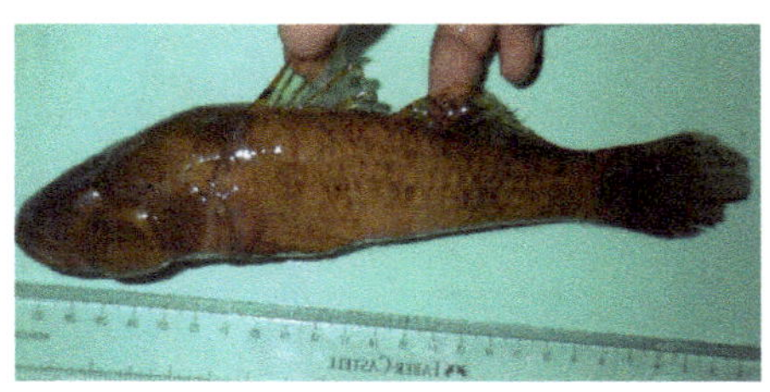

Glossogobius giuris

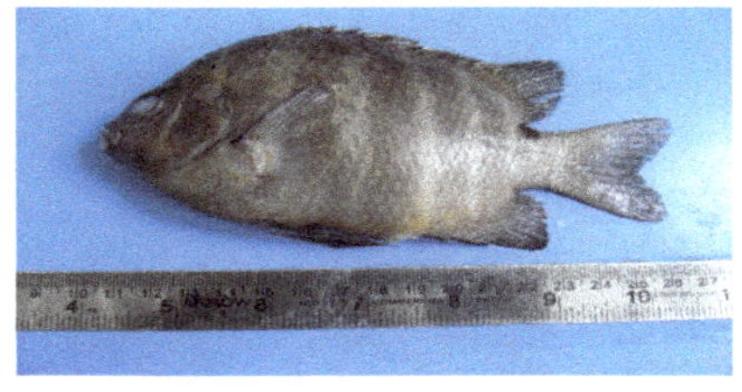

Abudefduf septemfasciatus (Cuvier, 1830)

Scatophagus argus (Linnaeus, 1766)

Siganus canaliculatus (Park, 1797)

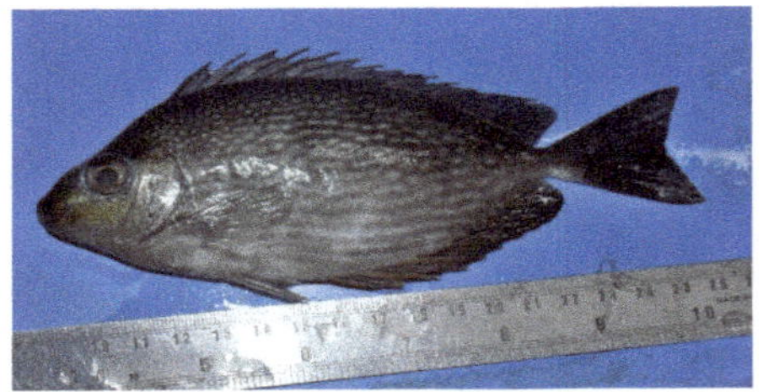

Siganus javus (Linnaeus, 1766)

Siganus lineatus (Valenciennes, 1835)

Sphyraena jello Cuvier, 1829

Channa punctata (Bloch, 1793)

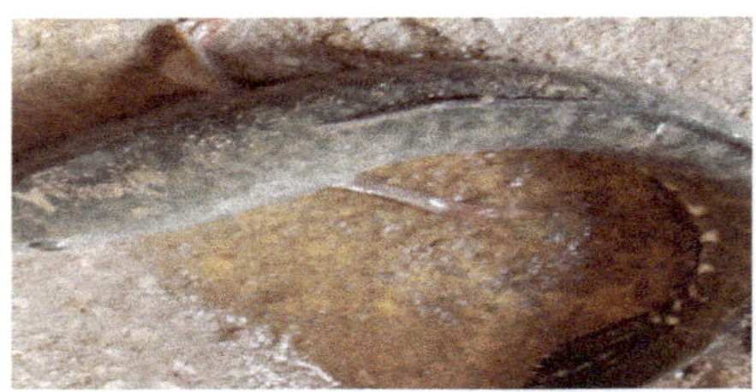

Channa striata (Bloch, 1793)

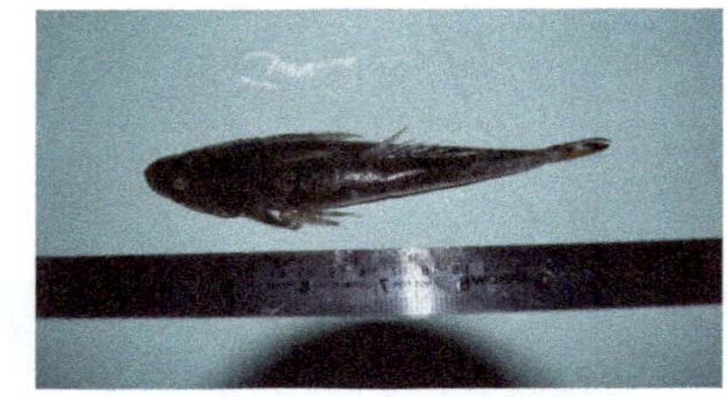

Platycephalus indicus (Linnaeus, 1758)

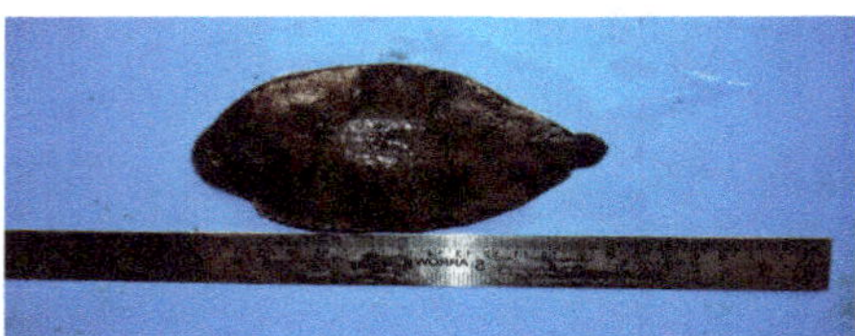

Brachirus orientalis (Bloch & Schneider, 1801)

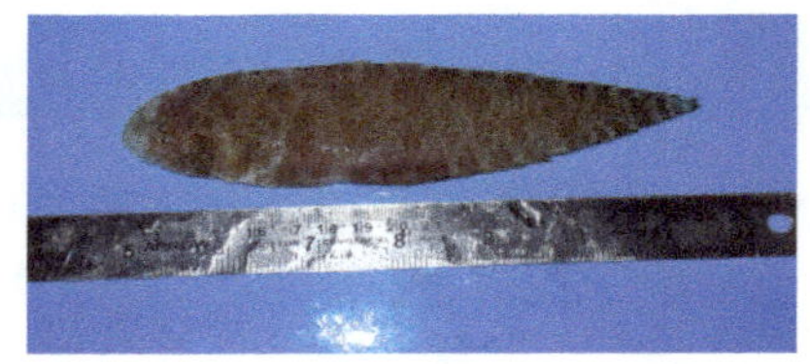

Cynoglossus arel (Bloch & Schneider, 1801)

Triacanthus biaculeatus (Bloch, 1786)

Arothron leopardus (Day, 1878)

www.ingramcontent.com/pod-product-compliance
Ingram Content Group UK Ltd.
Pitfield, Milton Keynes, MK11 3LW, UK
UKHW021010290726
14059UKWH00001BA/68